Internet of Things with Arduino Blueprints

Develop interactive Arduino-based Internet projects
with Ethernet and Wi-Fi

Pradeeka Seneviratne

PUBLISHING

BIRMINGHAM - MUMBAI

Internet of Things with Arduino Blueprints

First published: October 2015

Production reference: 1201015

Published by Packt Publishing Ltd.
Livery Place
35 Livery Street
Birmingham B3 2PB, UK.

ISBN 978-1-78528-548-6

www.packtpub.com

Credits

Author
Pradeeka Seneviratne

Reviewers
Francesco Azzola
Paul Deng
Charalampos Doukas
Paul Massey

Commissioning Editor
Nadeem Bagban

Acquisition Editor
Vivek Anantharaman

Content Development Editor
Arwa Manasawala

Technical Editor
Vivek Arora

Copy Editors
Imon Biswas
Angad Singh

Project Coordinator
Shweta H Birwatkar

Proofreader
Safis Editing

Indexer
Tejal Soni

Graphics
Jason Monteiro

Production Coordinator
Aparna Bhagat

Cover Work
Aparna Bhagat

About the Author

Pradeeka Seneviratne is a software engineer with over 10 years of experience in computer programming and systems designing. He loves programming embedded systems such as Arduino and Raspberry Pi. Pradeeka started learning about electronics when he was at primary college by reading and testing various electronic projects found in newspapers, magazines, and books.

Pradeeka is currently a full-time software engineer who works with highly scalable technologies. Previously, he worked as a software engineer for several IT infrastructure and technology servicing companies, and he was also a teacher for information technology and Arduino development.

He researches how to make Arduino-based unmanned aerial vehicles and Raspberry Pi-based security cameras.

About the Reviewers

Francesco Azzola is an electronics engineer with more than 15 years of experience in the architecture and development of JEE applications. He has a deep knowledge of mobile messaging, smart cards, and mobile applications. He enjoys building Android apps and experimenting with the IoT ecosystem using Arduino and Android. He is a Sun Certified Enterprise Architect (SCEA), SCWCD, SCJP, Prince2 (Foundation), and VCA-DCV. In his spare time, he runs a blog about Android and IoT (http://www.survivingwithandroid.com/).

Paul Deng is a senior software engineer with over 8 years of experience in IoT app design and development. He has been working with the Arduino platform since its early days in 2008.

His past experience involves end-to-end IoT app design and development, including embedded devices, large-scale machine learning, and cloud and web apps. Paul holds software algorithm patents and was a finalist of the Shell Australian Innovation Challenge 2011. He has authored several publications on IoT and sensor networks.

Paul is an open source contributor and active blogger. He is also an AWS Certified Solutions Architect and Developer with a master's degree in distributed computing from the University of Melbourne.

He lives in Melbourne, Australia with his wife, Cindy, and son, Leon. You can visit his website at http://dengpeng.de/ to see what he is currently exploring and to learn more about him.

Paul Massey has worked in computer programming for over 20 years, 11 years of which have been as a CEO of Scriptwerx (http://ghost.scriptwerx.io/). He is an expert in JavaScript and mobile technologies, as well as working with the Arduino platform (and similar platforms). He has worked on this platform for a number of years, creating hardware and software projects for Internet of Things, audio-visual, and automotive technologies.

www.PacktPub.com

Support files, eBooks, discount offers, and more

For support files and downloads related to your book, please visit www.PacktPub.com.

Did you know that Packt offers eBook versions of every book published, with PDF and ePub files available? You can upgrade to the eBook version at www.PacktPub.com and as a print book customer, you are entitled to a discount on the eBook copy. Get in touch with us at service@packtpub.com for more details.

At www.PacktPub.com, you can also read a collection of free technical articles, sign up for a range of free newsletters and receive exclusive discounts and offers on Packt books and eBooks.

https://www2.packtpub.com/books/subscription/packtlib

Do you need instant solutions to your IT questions? PacktLib is Packt's online digital book library. Here, you can search, access, and read Packt's entire library of books.

Why subscribe?

- Fully searchable across every book published by Packt
- Copy and paste, print, and bookmark content
- On demand and accessible via a web browser

Free access for Packt account holders

If you have an account with Packt at www.PacktPub.com, you can use this to access PacktLib today and view 9 entirely free books. Simply use your login credentials for immediate access.

Table of Contents

Preface

Arduino is a small single-chip computer board that can be used for a wide variety of creative hardware projects. The hardware consists of a simple microcontroller, board, and chipset. It comes with a Java-based IDE that allows creators to program the board. Arduino is the ideal open hardware platform to experiment with the world of Internet of Things. The credit card-sized Arduino board can be used via the Internet to make useful and interactive Internet of Things (IoT) projects.

Internet of Things with Arduino Blueprints is a project-based book that begins with projects based on IoT and cloud computing concepts. This book covers up to eight projects that will allow devices to communicate with each other, access information over the Internet, store and retrieve data, and interact with users — creating smart, pervasive, and always connected environments. It explains how wired and wireless Internet connections can be used with projects and explains the use of various sensors and actuators. The main aim of this book is to teach you how Arduino can be used for Internet-related projects so that users are able to control actuators, gather data from various kinds of sensors, and send and receive data wirelessly across HTTP and TCP protocols.

Finally, you can use these projects as blueprints for many other IoT projects and put them to good use. By the end of the book, you will be an expert in the use of IoT with Arduino to develop a set of projects that can relate very well to IoT applications in the real world.

What this book covers

Chapter 1, Internet-Controlled PowerSwitch, briefly introduces Arduino UNO and Arduino Ethernet shield basics while focusing on building an Internet-controlled PowerSwitch using Arduino UNO, the Arduino Ethernet shield, and PowerSwitch Tail to turn electrical appliances on/off through the Internet via a web-based user interface. Also, you will learn how to increase the complexity of PowerSwitch by adding a circuit to track the mains electricity.

Chapter 2, Wi-Fi Signal Strength Reader and Haptic Feedback, briefly introduces Arduino Wi-Fi shield basics, vibration motors, and haptic feedback. You will learn how to make advanced vibration patterns using vibration motors with a haptic motor controller and the Adafruit haptic library according to the Wi-Fi signal strength received by the Arduino Wi-Fi shield.

Chapter 3, Internet-Connected Smart Water Meter, focuses on building a flow sensor-based water meter in conjunction with the Arduino Ethernet shield to measure water flow rate and volume, and then display them on an LCD screen. In addition, you will learn how to convert this water meter to a web server and request readings through the Internet of Ethernet.

Chapter 4, Arduino Security Camera with Motion Detection, explains how to incrementally develop a Arduino Ethernet shield-based security camera with the Adafruit TTL Serial JPEG camera and the VC0706 camera library. In addition, you will learn how to add motion detection functionality and upload the captured images to Flickr.

Chapter 5, Solar Panel Voltage Logging with NearBus Cloud Connector and Xively, briefly introduces the NearBus cloud connector and Xively, while focusing on building a solar panel voltage logger with the Arduino Ethernet shield with a few electronic components. Also, you will learn how to log the output voltage of a solar panel in combination with NearBus and Xively, and then display real-time data that can be viewed through a web browser.

Chapter 6, GPS Location Tracker with Temboo, Twilio, and Google Maps, briefly introduces the GPS shield and how to use the TinyGPSPlus library and the Google JavaScript API library to build a real-time location tracker to display the current location of the GPS shield on Google Maps. You will also learn the basics of Temboo and Twilio cloud services.

Chapter 7, Tweet-a-Light – Twitter-Enabled Electric Light, introduces Python, a more powerful programming language that can be used to read Twitter tweets and write data to a computer's serial port accordingly. Finally, you will learn to build an electric light switch that can be controlled using Twitter tweets to turn the switch on and off.

Chapter 8, Controlling Infrared Devices Using IR Remote, focuses on building an infrared remote control with the Arduino Ethernet shield and a few electronic components that can be controlled through the Ethernet or Internet to control IR-enabled devices remotely. You will learn how to record and reproduce IR commands using the Arduino IR remote library. In addition, you will learn how to add IR functionality to non-IR enabled devices.

What you need for this book

This book has been written and tested on the Windows environment and uses various software components with Arduino. It would be great if you could prepare your development environment before proceeding with the sample code provided along with each chapter. The following list briefly gives you the details about the software requirements that you should have to set up your PC for each chapter:

- The Arduino software: This is the main development environment that you will use to write, verify, and run your sketches in every chapter of this book. The latest Arduino installer for Windows can be downloaded from https://www.arduino.cc/en/Main/Software. Throughout this book, we will write and test Arduino sketches in the Windows environment.

- A web browser: Normally, every PC has a default web browser, such as Microsoft Internet Explorer (or Microsoft Edge in Windows 10), Google Chrome, or Mozilla Firefox.

- The Adafruit DRV2605 library: You need this library to control vibrators (that is, vibration motors) with the Adafruit DRV2605 haptic controller in *Chapter 2, Wi-Fi Signal Strength Reader and Haptic Feedback*. You can download this library from https://github.com/adafruit/Adafruit_DRV2605_Library.

- The Adafruit VC0706 camera library: You will need this library to interface the Adafruit VC0706 Serial JPEG camera with Arduino in *Chapter 4, Arduino Security Camera with Motion Detection*. You can download this library from https://github.com/adafruit/Adafruit-VC0706-Serial-Camera-Library.

- NearBus Agent (An Arduino library for Ethernet): You will need this library to connect the Arduino Ethernet shield with the NearBus cloud connector for Arduino memory mapping with NearBus in *Chapter 5, Solar Panel Voltage Logging with NearBus Cloud Connector and Xively*. You can download this library from http://www.nearbus.net/downloads/NearBusEther_v16.zip.

- FlexiTimer2: This will make sure that Arduino correctly functions with the NearBus Agent library in *Chapter 5, Solar Panel Voltage Logging with NearBus Cloud Connector and Xively*. You can download it from http://github.com/wimleers/flexitimer2/zipball/v1.1.

- The TinyGPSPlus library: This will be required to work with the SparkFun GPS shield in *Chapter 6, GPS Location Tracker with Temboo, Twilio, and Google Maps* and can be downloaded from https://github.com/mikalhart/TinyGPSPlus/archive/master.zip.

- Python: You can download Python from `https://www.python.org/`, and the instructions about the download and installation can be found in *Chapter 7, Tweet-a-Light – Twitter-Enabled Electric Light*.

- The Arduino IR remote library: You will need this library to send and receive and extract IR commands in *Chapter 8, Controlling Infrared Devices Using IR Remote*. You can download it from `https://github.com/z3t0/Arduino-IRremote`.

Who this book is for

This book is intended for those who want to learn more about Arduino and make Internet-based interactive projects with Arduino. If you are an experienced software developer who understands the basics of electronics, then you can quickly learn how to build the Arduino projects explained in this book.

Conventions

In this book, you will find a number of text styles that distinguish between different kinds of information. Here are some examples of these styles and an explanation of their meaning.

Code words in text, database table names, folder names, filenames, file extensions, pathnames, dummy URLs, user input, and Twitter handles are shown as follows: "Copy the following `index.html` file from the code folder of `Chapter 5` to your computer's hard drive."

A block of code is set as follows:

```
// Set the picture size - you can choose one of 640x480, 320x240 or
160x120
// Remember that bigger pictures take longer to transmit!
cam.setImageSize(VC0706_640x480);  // biggest
//cam.setImageSize(VC0706_320x240);  // medium
//cam.setImageSize(VC0706_160x120);   // small
```

When we wish to draw your attention to a particular part of a code block, the relevant lines or items are set in bold:

```
int32_t WiFiDrv::getCurrentRSSI()
{
  startScanNetworks();
  WAIT_FOR_SLAVE_SELECT();
  // Send Command
  SpiDrv::sendCmd(GET_CURR_RSSI_CMD, PARAM_NUMS_1);
```

Any command-line input or output is written as follows:

```
>python your_python_script.py
```

New terms and **important words** are shown in bold. Words that you see on the screen, for example, in menus or dialog boxes, appear in the text like this: "Click on the **Create an App** link if it is not selected by default."

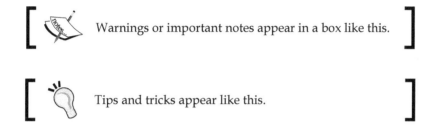

> Warnings or important notes appear in a box like this.

> Tips and tricks appear like this.

Reader feedback

Feedback from our readers is always welcome. Let us know what you think about this book—what you liked or disliked. Reader feedback is important for us as it helps us develop titles that you will really get the most out of.

To send us general feedback, simply e-mail feedback@packtpub.com, and mention the book's title in the subject of your message.

If there is a topic that you have expertise in and you are interested in either writing or contributing to a book, see our author guide at www.packtpub.com/authors.

Customer support

Now that you are the proud owner of a Packt book, we have a number of things to help you to get the most from your purchase.

Downloading the example code

You can download the example code files from your account at http://www.packtpub.com for all the Packt Publishing books you have purchased. If you purchased this book elsewhere, you can visit http://www.packtpub.com/support and register to have the files e-mailed directly to you.

Downloading the color images of this book

We also provide you with a PDF file that has color images of the screenshots/ diagrams used in this book. The color images will help you better understand the changes in the output. You can download this file from: `https://www. packtpub.com/sites/default/files/downloads/5486OS_ColoredImages.pdf`.

Errata

Although we have taken every care to ensure the accuracy of our content, mistakes do happen. If you find a mistake in one of our books—maybe a mistake in the text or the code—we would be grateful if you could report this to us. By doing so, you can save other readers from frustration and help us improve subsequent versions of this book. If you find any errata, please report them by visiting `http://www.packtpub. com/submit-errata`, selecting your book, clicking on the **Errata Submission Form** link, and entering the details of your errata. Once your errata are verified, your submission will be accepted and the errata will be uploaded to our website or added to any list of existing errata under the Errata section of that title.

To view the previously submitted errata, go to `https://www.packtpub.com/books/ content/support` and enter the name of the book in the search field. The required information will appear under the **Errata** section.

Piracy

Piracy of copyrighted material on the Internet is an ongoing problem across all media. At Packt, we take the protection of our copyright and licenses very seriously. If you come across any illegal copies of our works in any form on the Internet, please provide us with the location address or website name immediately so that we can pursue a remedy.

Please contact us at `copyright@packtpub.com` with a link to the suspected pirated material.

We appreciate your help in protecting our authors and our ability to bring you valuable content.

Questions

If you have a problem with any aspect of this book, you can contact us at `questions@packtpub.com`, and we will do our best to address the problem.

1
Internet-Controlled PowerSwitch

For many years, people physically interacted with electrical appliances using hardware switches. Now that things have changed, thanks to the advances in technology and hardware, controlling a switch over the Internet without any form of physical interaction has become possible.

In this chapter, we will incrementally build a web server-enabled smart power switch that can be controlled through the Internet with a wired Internet connection. Let's move to Arduino's **IoT (Internet of Things)**.

In this chapter, you will do the following:

* Learn about Arduino UNO and Arduino Ethernet Shield basics
* Learn how to connect a PowerSwitch Tail with Arduino UNO
* Build a simple web server to handle client requests and control the PowerSwitch accordingly
* Build a simple mains electricity (general purpose alternating current) sensor with 5V DC wall power supply
* Develop a user friendly **UI (User Interface)** with **HTML (Hyper Text Markup Language)** and Metro UI **CSS (Cascade Style Sheet)**

Getting started

This project consists of a **DC** (**Direct Current**) activated relay switch with an embedded web server that can be controlled and monitored through the Internet and the integrated mains electricity sensor that can be used to get the status of the availability of mains electricity. The possible applications are:

- Controlling electrical devices such as lamp posts, water pumps, gates, doors, and so on, in remote locations

- Sensing the availability of mains electricity in houses, offices, and factories remotely

- Detecting whether a door, window, or gate is open or shut

Hardware and software requirements

All the hardware and software requirements are mentioned within each experiment. Most of the hardware used in this project are open source, which allows you to freely learn and hack them to make more creative projects based on the blueprints of this chapter.

Arduino Ethernet Shield

Arduino Ethernet Shield is used to connect your Arduino UNO board to the Internet. It is an open source piece of hardware and is exactly the same size as the Arduino UNO board. The latest version of the Arduino Ethernet Shield is **R3** (**Revision 3**). The official Arduino Ethernet Shield is currently manufactured in Italy and can be ordered through the official Arduino website (`https://store.arduino.cc`). Also, there are many Arduino Ethernet Shield clones manufactured around the world that may be cheaper than the official Arduino Ethernet Shield. This project is fully tested with a clone of Arduino Ethernet Shield manufactured in China.

Arduino UNO R3 (Front View)

Arduino Ethernet Shield R3 (Front View)

Plug your Arduino Ethernet Shield into your Arduino UNO board using wire wrap headers so that it's exactly intact with the pin layout of the Arduino UNO board. The following image shows a stacked Arduino UNO and Arduino Ethernet Shield together:

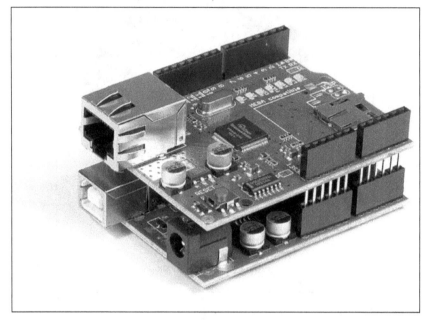

Arduino Ethernet Shield R3 (top) is stacked with Arduino UNO R3 (bottom) (Front View)

Arduino Ethernet Shield consists of an Ethernet controller chip—WIZnet W5100— the only proprietary hardware used with the shield. The WIZnet W5100 includes a fully hardwired TCP/IP stack, integrated Ethernet **MAC** (**Media Access Control**), and **PHY** (**Physical Layer**).

The hardwired TCP/IP stack supports the following protocols:

- **TCP (Transport Control Protocol)**
- **UDP (User Datagram Protocol)**
- **IPv4 (Internet Protocol Version 4)**
- **ICMP (Internet Control Message Protocol)**
- **ARP (Address Resolution Protocol)**
- **IGMP (Internet Group Management Protocol)**
- **PPPoE (Point-to-Point Protocol over Ethernet)**

The WIZnet W5100 Ethernet controller chip also simplifies the Internet connectivity without using an operating system.

The WIZnet W5100 Ethernet controller (Top View)

Throughout this chapter, we will only work with TCP and IPv4 protocols.

The Arduino UNO board communicates with the Arduino Ethernet Shield using digital pins 10, 11, 12, and 13. Therefore, we will not use these pins in our projects to make any external connections. Also, digital pin 4 is used to select the SD card that is installed on the Arduino Ethernet Shield, and digital pin 10 is used to select the Ethernet controller chip. This is called **SS (Slave Select)** because the Arduino Ethernet Shield is acting as the slave and the Arduino UNO board is acting as the master.

However, if you want to disable the SD card and use digital pin 4, or disable the Ethernet controller chip and use digital pin 10 with your projects, use the following code snippets inside the `setup()` function:

1. To disable the SD card:

    ```
    pinMode(4,OUTPUT);
    digitalWrite(4, HIGH);
    ```

2. To disable the Ethernet Controller chip:

    ```
    pinMode(10,OUTPUT);
    digitalWrite(10, HIGH);
    ```

The Arduino Ethernet board

The Arduino Ethernet board is a new version of the Arduino development board with the WIZnet Ethernet controller built into the same board. The USB to serial driver is removed from the board to keep the board size the same as Arduino UNO and so that it can be stacked with any Arduino UNO compatible shields on it.

You need an FTDI cable compatible with 5V to connect and program your Arduino Ethernet board with a computer.

The Arduino Ethernet board (Front View)

FTDI cable 5V (Source: https://commons.wikimedia.org/wiki/File:FTDI_Cable.jpg)

You can visit the following links to get more information about the Arduino Ethernet board and FTDI cable:

- The Arduino Ethernet board (`https://store.arduino.cc/product/A000068`)

- FTDI cable (`https://www.sparkfun.com/products/9717`)

You can build all the projects that are explained within this chapter and other chapters throughout the book with the Arduino Ethernet board using the same pin layout.

Connecting Arduino Ethernet Shield to the Internet

To connect your Ethernet shield to the Internet, you require the following hardware:

- An Arduino UNO R3 board (`https://store.arduino.cc/product/A000066`)

- A 9VDC 650mA wall adapter power supply. The barrel connector of the power supply should be center positive 5.5 x 2.1 mm. (Here is the link for a perfect fit: `https://www.sparkfun.com/products/298`)

- A USB A-to-B male/male-type cable. These types of cables are usually used for printers (`https://www.sparkfun.com/products/512`)

- A Category 6 Ethernet cable (`https://www.sparkfun.com/products/8915`)

- A router or switch with an Internet connection

Use the following steps to make connections between each hardware component:

1. Plug your Ethernet shield into your Arduino board using soldered wire wrap headers:

Fritzing representation of Arduino and Ethernet shield stack

2. Get the Ethernet cable and connect one end to the Ethernet jack of the Arduino Ethernet Shield.

One end of the Ethernet cable is connected to the Arduino Ethernet board

3. Connect the other end of the Ethernet cable to the Ethernet jack of the network router or switch.

The other end of the Ethernet cable is connected to the router/switch

4. Connect the 9VDC wall adapter power supply to the DC barrel connector of the Arduino board.

5. Use the USB A-to-B cable to connect your Arduino board to the computer. Connect the type A plug end to the computer and the type B plug end to the Arduino board.

One end of the Ethernet cable is connected to the Ethernet shield (top) and the power connector and USB cable are connected to the Arduino board (bottom) Image courtesy of SparkFun Electronics (https://www.sparkfun.com)

Testing your Arduino Ethernet Shield

To test you Arduino Ethernet Shield, follow these steps:

1. Open your Arduino IDE and navigate to **File | Examples | Ethernet | WebServer**:

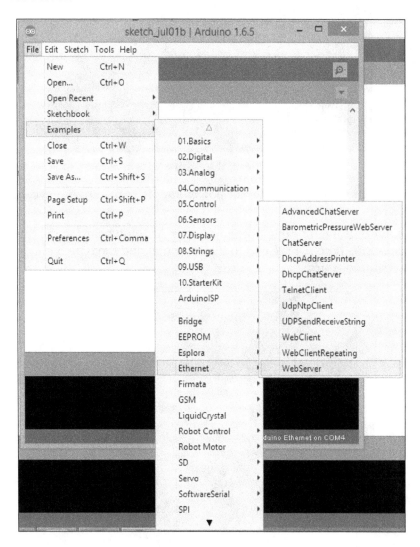

2. The sample sketch **WebServer** will open in a new Arduino IDE:

3. You can also paste the code from the sketch named `B04844_01_01.ino` from the code folder of this chapter. The following header files should be included for serial communication and Ethernet communication in the beginning of the sketch:

```
#include <SPI.h> //initiates Serial Peripheral Interface
#include <Ethernet.h> //initiates Arduino Ethernet library
```

4. Replace the MAC address with your Ethernet shield's MAC address if you know it. You can find the printed sticker of the MAC address affixed to the back of your Ethernet shield. (Some clones of Arduino Ethernet Shield don't ship with a MAC address affixed on them). If you don't know the MAC address of your Arduino Ethernet Shield, use the one mentioned in the sample code or replace it with a random one. But don't use network devices with the same MAC address on your network; it will cause conflicts and your Ethernet shield will not function correctly. (Read *Finding the MAC address and obtaining a valid IP address* for more information on MAC addresses).

```
byte mac[] = {0xDE, 0xAD, 0xBE, 0xEF, 0xFE, 0xED};
```

5. Replace the IP address with a static IP in your local network IP range. (Read the *Finding the MAC address and obtaining a valid IP address* section for selecting a valid IP address).

```
IPAddress ip(192, 168, 1, 177);
```

6. Then, create an instance of the Arduino Ethernet Server library and assign port number 80 to listen to incoming HTTP requests.

```
EthernetServer server(80);
```

7. Inside the `setup()` function, open the serial communications and wait for the port to open. The computer will communicate with Arduino at a speed of 9600 bps.

```
Serial.begin(9600);
```

8. The following code block will start the Ethernet connection by using the MAC address and IP address (we have assigned a static IP address) of the Arduino Ethernet Shield and start the server. Then it will print the IP address of the server on Arduino Serial Monitor using `Ethernet.localIP()`:

```
Ethernet.begin(mac, ip);
server.begin();
Serial.print("server is at ");
Serial.println(Ethernet.localIP());
```

9. Inside the `loop()` function, the server will listen for incoming clients.

```
EthernetClient client = server.available();
```

10. If a client is available, the server will connect with the client and read the incoming HTTP request. Then, reply to the client by the standard HTTP response header. The output can be added to the response header using the `EthernetClient` class's `println()` method:

```
if (client) {
  Serial.println("new client");
  // an http request ends with a blank line
  boolean currentLineIsBlank = true;
  while (client.connected()) {
    if (client.available()) {
      char c = client.read();
      Serial.write(c);
      // if you've gotten to the end of the line
      (received a newline
      // character) and the line is blank, the http
      request has ended,
      // so you can send a reply
      if (c == '\n' && currentLineIsBlank) {
```

```
        // send a standard http response header
        client.println("HTTP/1.1 200 OK");
        client.println("Content-Type: text/html");
        client.println("Connection: close");  // the
        connection will be closed after completion of the
        response
        client.println("Refresh: 5");  // refresh the
        page automatically every 5 sec
        client.println();
        client.println("<!DOCTYPE HTML>");
        client.println("<html>");
        // output the value of each analog input pin
        for (int analogChannel = 0; analogChannel < 6;
        analogChannel++) {
          int sensorReading = analogRead(analogChannel);
          client.print("analog input ");
          client.print(analogChannel);
          client.print(" is ");
          client.print(sensorReading);
          client.println("<br />");
        }
        client.println("</html>");
        break;
      }
      if (c == '\n') {
        // you're starting a new line
        currentLineIsBlank = true;
      }
      else if (c != '\r') {
        // you've gotten a character on the current line
        currentLineIsBlank = false;
      }
    }
  }
  // give the web browser time to receive the data
  delay(1);
```

11. Finally, close the connection from the client using the `EthernetClient` class's `stop()` method:

```
client.stop();
Serial.println("client disconnected");
```

12. Verify the sketch by clicking on the **Verify** button located in the toolbar.

13. On the menu bar, select the board by navigating to **Tools | Board | Arduino UNO**. If you are using an Arduino Ethernet board, select **Tools | Board | Arduino Ethernet**.

14. On the menu bar, select the **COM** port by navigating to **Tools | Port** and then selecting the port number.

15. Upload the sketch into your Arduino UNO board by clicking on the **Upload** button located in the toolbar.

16. Open your Internet browser (such as Google Chrome, Mozilla Firefox, or Microsoft Internet Explorer) and type the IP address (`http://192.168.1.177/`) assigned to your Arduino Ethernet Shield in the sketch (in Step 4), and hit the *Enter* key.

17. The web browser will display analog input values (impedance) of all the six analog input pins (A0-A5). The browser will refresh every 5 seconds with the new values. Use following code to change the automatic refresh time in seconds:

```
client.println("Refresh: 5");
```

Output for Arduino Ethernet board: Analog input values are displaying on the Google Chrome browser, refreshing every 5 seconds

Output for Arduino UNO + Arduino Ethernet Shield: Analog input values are displaying on the Google Chrome browser, refreshing every 5 seconds

Arduino Serial Monitor prints the static IP address of Arduino Ethernet Shield

18. To make your sketch more stable and to ensure that it does not hang, you can do one of the following:

   ° Remove the SD card from the slot.

   ° Add the following two lines inside your `setup()` function:

   ```
   pinMode(4,OUTPUT);
   digitalWrite(4, HIGH);
   ```

Now you can be assured that your Arduino Ethernet Shield is working properly and can be accessed through the Internet.

Selecting a PowerSwitch Tail

PowerSwitch Tail has a built-in AC relay that is activated between 3-12 VDC. This is designed to easily integrate with many microcontroller platforms, such as Arduino, Raspberry Pi, BeagleBone, and so on. Usually, Arduino digital output provides 5VDC that allows it to activate the **AC** (**Alternative Current**) relay inside the PowerSwitch Tail. Using a PowerSwitch Tail with your microcontroller projects provides safety since it distinguishes between AC and DC circuitry by using an optocoupler which is an optically activated switch.

PowerSwitch Tail ships in several variants. At the time of writing this book, the product website lists various PowerSwitch Tails, assembled and in kit form, that can be used with this project.

To build this project, we will use a 240V AC PowerSwitch Tail that can be purchased as a kit and assembled.

PN PSSRKT-240

Refer to `http://www.powerswitchtail.com/Pages/PowerSwitchTail240vackit.aspx`.

PN PSSRKT-240 Normally Open (NO) version—240V AC Image courtesy of PowerSwitchTail.com, LLC (`http://www.powerswitchtail.com`)

Here, we will not cover the assembly instructions about the PN PSSRKT-240 kit. However, you can find the assembly instructions at `http://www.powerswitchtail.com/Documents/PSSRTK%20Instructions.pdf`.

The following image shows an assembled PN PSSRKT-240 kit:

PN PSSRKT-240 Normally Open (NO) version—240V Image courtesy of PowerSwitchTail.com, LLC (`http://www.powerswitchtail.com`)

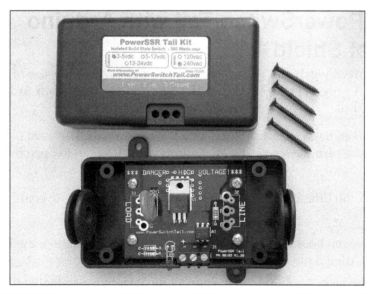

PN PSSRKT-240 Normally Open (NO) version—240V Image courtesy of PowerSwitchTail.com, LLC (`http://www.powerswitchtail.com`)

If you are in a country that has a 120V AC connection, you can purchase an assembled version of the PowerSwitch Tail.

PN80135

Refer to http://www.powerswitchtail.com/Pages/default.aspx.

PN80135 Normally Open (NO) version—120V AC (left-hand side plug for LOAD and right-hand side plug for LINE) Image courtesy of SparkFun Electronics (https://www.sparkfun.com)

Wiring PowerSwitch Tail with Arduino Ethernet Shield

Wiring the PowerSwitch Tail with Arduino is very easy. Use any size of wire range between gauge #14-30 AWG to make the connection between Arduino and PowerSwitch Tail.

PowerSwitch Tail has a terminal block with three terminals. Use a small flat screwdriver and turn the screws **CCW (Counter Clock Wise)** to open the terminal contacts.

With the Arduino Ethernet Shield mounted on the Arduino UNO board, do the following:

1. Use the red hookup wire to connect the positive terminal of the PowerSwitch Tail to digital pin 5 on your Arduino.

2. Use the black hookup wire to connect the negative terminal of the PowerSwitch Tail to the GND pin on your Arduino.

3. Connect the wall adapter power supply (9V DC 650mA) to the DC power jack on your Arduino board. The ground terminal is connected internally to the AC-side electrical safety ground (the green conductor) and can be used if needed.

Two wires from Arduino connected to the PowerSwitch Tail Image courtesy of PowerSwitchTail.com, LLC (http://www.powerswitchtail.com)

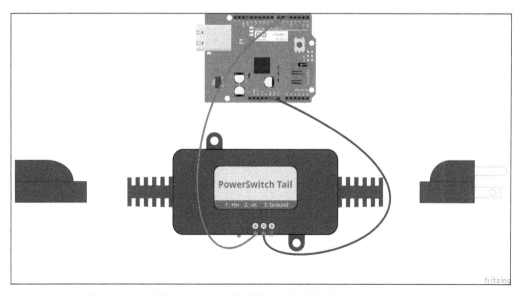

PowerSwitch Tail connected to the Ethernet Shield—Fritzing representation

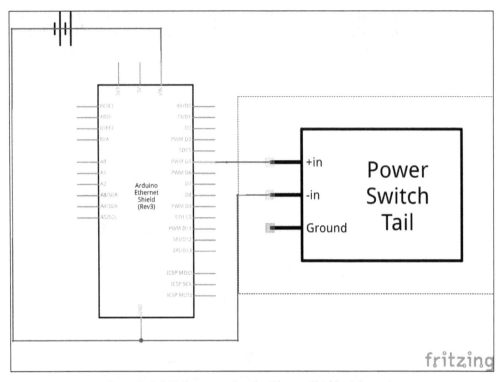

PowerSwitch Tail connected to the Ethernet Shield—Schematic

Turning PowerSwitch Tail into a simple web server

In this topic, we will look into how to convert our Arduino connected PowerSwitch Tail into a simple web server to handle client requests, such as the following:

- Turn ON the PowerSwitch Tail
- Turn OFF the PowerSwitch Tail

And other useful information such as:

- Display the current status of the PowerSwitch Tail
- Display the presence or absence of the main electrical power

What is a web server?

A web server is a piece of software which serves to connected clients. An Arduino web server uses HTTP on top of TCP and UDP. But remember, the Arduino web server can't be used as a replacement for any web server software running on a computer because of the lack of processing power and limited number of multiple client connectivity.

A step-by-step process for building a web-based control panel

In this section ,you will learn how to build a web-based control panel for controlling the PowerSwitch Tail through the Internet.

We will use the Arduino programming language and HTML that's running on the Arduino web server. Later, we will add HTML radio button controls to control the power switch.

Handling client requests by HTTP GET

Using the HTTP GET method, you can send a query string to the server along with the URL.

The query string consists of a name/value pair. The query string is appended to the end of the URL and the syntax is `http://example.com?name1=value1`.

Also, you can add more name/value pairs to the URL by separating them with the & character, as shown in the following example:

`http://example.com?name1=value1&name2=value2`.

So, our Arduino web server can actuate the PowerSwitch Tail using the following URLs:

- To turn ON the PowerSwitch Tail: `http://192.168.1.177/?switch=1`
- To turn OFF the PowerSwitch Tail: `http://192.168.1.177/?switch=0`

The following sketch can be used by the web server to read the incoming client requests, process them, and actuate the relay inside the PowerSwitch Tail:

1. Open your Arduino IDE and type or paste the code from the `B04844_01_02.ino` sketch.

2. In the sketch, replace the MAC address with your Arduino Ethernet Shield's MAC address:

```
byte mac[] = { 0x90, 0xA2, 0xDA, 0x0B, 0x00 and 0xDD };
```

3. Replace the IP address with an IP valid static IP address in the range of your local network:

```
IPAddress ip(192,168,1,177);
```

4. If you want the IP address dynamically assigned by the DHCP to the Arduino Ethernet Shield, do the following:

    1. Comment the following line in the code:

    ```
    //IPAddress ip(192,168,1,177);
    ```

    2. Comment the following line in the code:

    ```
    //Ethernet.begin(mac, ip);
    ```

    3. Uncomment the following line in the code:

    ```
    Ethernet.begin(mac);
    ```

5. The following two lines will read the incoming HTTP request from the client using the `EthernetClient` class's `read()` method and store it in a string variable `http_Request`:

```
char c = client.read();
http_Request += c;
```

6. The following code snippet will check whether the HTTP request string contains the query string that is sent to the URL. If found, it will turn on or off the PowerSwitch Tail according to the name/value pair logically checked inside the sketch.

    The `indexOf()` function can be used to search for the string within another string. If it finds the string `switch=1` inside the HTTP request string, the Arduino board will turn digital pin 5 to the HIGH state and turn on the PowerSwitch Tail. If it finds the text `switch=0`, the Arduino board will turn the digital pin 5 to the LOW state and turn off the PowerSwitch Tail.

```
if (httpRequest.indexOf("GET /?switch=0 HTTP/1.1") > -1) {
            relayStatus = 0;
            digitalWrite(5, LOW);
```

```
Serial.println("Switch is Off");
} else if (httpRequest.indexOf("GET /?switch=1
HTTP/1.1") > -1) {
  relayStatus = 1;
  digitalWrite(5, HIGH);
  Serial.println("Switch is On");
}
```

7. Select the correct Arduino board and COM port from the menu bar.

8. Verify and upload the sketch into your Arduino UNO board (or the Arduino Ethernet board).

9. If you have to choose DHCP to assign an IP address to your Arduino Ethernet Shield, it will be displayed on the Arduino Serial Monitor. On the menu bar, go to **Tools** | **Serial Monitor**. The Arduino Serial Monitor window will be displayed with the IP address assigned by the DHCP.

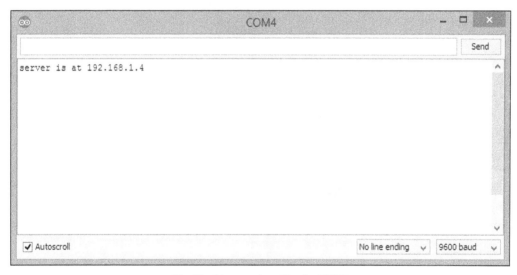

The IP address assigned by the DHCP

10. Plug the PowerSwitch Tail LINE side into the wall power socket and connect the lamp into the LOAD side of the PowerSwitch Tail. Make sure that the lamp switch is in the ON position and all the switches of the wall power socket are in the ON position.

11. Open your Internet browser and type the IP address of your Arduino Ethernet Shield with HTTP protocol. For our example it is `http://192.168.1.177`. Then hit the *Enter* key on your keyboard.

12. The web browser sends an HTTP request to the Arduino web server and the web server returns the processed web content to the web browser. The following screen capture displays the output in the web browser.

13. Type `http://192.168.1.177/?switch=1` and hit the *Enter* key. The lamp will turn on.

14. Type `http://192.168.1.177/?switch=0` and hit the *Enter* key. The lamp will turn off.

15. If you have connected your Arduino Ethernet Shield to your home wireless network, you can test your PowerSwitch Tail using your Wi-Fi connected smartphone as well. If you have the idea to add port forwarding to your router, you can then control your switch from anywhere in the world. Explaining about port forwarding is out of scope of this book.

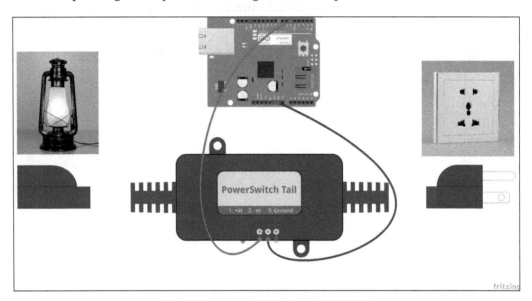

Electric lamp controlled by PowerSwitch Tail

PowerSwitch Tail control panel accessed by Google Chrome browser

Sensing the availability of mains electricity

You can sense the availability of mains electricity in your home and read the status before actuating the PowerSwitch Tail.

You will need the following hardware to build the sensor:

- A 5VDC 2A wall adapter power supply (https://www.sparkfun.com/products/12889)
- A 10 kilo Ohm resistor (https://www.sparkfun.com/products/8374)

Follow the next steps to attach the sensor to the Arduino Ethernet Shield:

1. Connect the positive wire of the 5V DC wall adapter power supply to the Ethernet shield digital pin 2.
2. Connect the negative wire of the wall adapter power supply to the Ethernet shield GND pin.
3. Connect the 10 kilo ohm resistor between the Ethernet shield digital pin 2 and the GND pin.

4. Plug the wall adapter power supply into the wall.

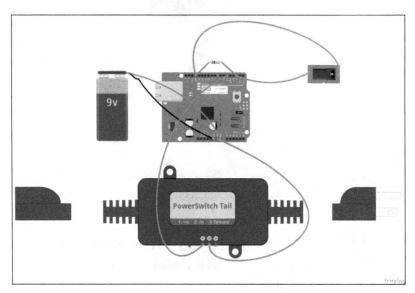

A wiring diagram

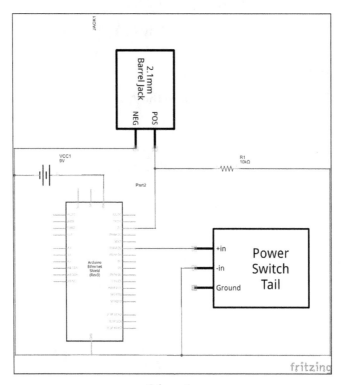

Schematic

Testing the mains electricity sensor

The previous sketch is modified to check the availability of the mains electricity and operate PowerSwitch Tail accordingly. The 5V DC wall adapter power supply plugged into the wall keeps the Arduino digital pin 2 in the HIGH state if mains electricity is available. If mains electricity is not available, the digital pin 2 switches to the LOW state.

1. Open your Arduino IDE and paste the code from the sketch named `B04844_01_03.ino` from the code folder of this chapter.

2. Power up your Arduino Ethernet Shield with 9V battery pack so that it will work even without mains electricity.

3. The Arduino digital pin 2 is in its HIGH state if mains electricity is available. The `hasElectricity` boolean variable holds the state of availability of the electricity.

4. If only the mains electricity is available, the PowerSwitch Tail can be turned ON. If not, the PowerSwitch Tail is already in its OFF state.

Building a user-friendly web user interface

The following Arduino sketch adds two radio buttons to the web page so the user can easily control the switch without typing the URL with the query string into the address bar of the web browser. The radio buttons will dynamically build the URL with the query string depending on the user selection and send it to the Arduino web server with the HTTP request.

1. Open your Arduino IDE and paste the code from the sketch named `B04844_01_04.ino` from the code folder of this chapter.

2. Replace the IP address with a new IP address in your local area network's IP address range.

   `IPAddress ip(192,168,1,177);`

3. Verify and upload the sketch on your Arduino UNO board.

4. Open your web browser and type your Arduino Ethernet Shield's IP address into the address bar and hit the *Enter* key.

5. The following code snippet will submit your radio button selection to the Arduino web sever as an HTTP request using the HTTP GET method. The radio button group is rendered inside the `<form method="get"></form>` tags.

```
client.println("<form method=\"get\">");
if (httpRequest.indexOf("GET /?switch=0
HTTP/1.1") > -1) {
   relayStatus = 0;
   digitalWrite(9, LOW);
   Serial.println("Off Clicked");
} else if (httpRequest.indexOf("GET /?switch=1
HTTP/1.1") > -1) {
   relayStatus = 1;
   digitalWrite(9, HIGH);
   Serial.println("On Clicked");
}

if (relayStatus) {
   client.println("<input type=\"radio\"
   name=\"switch\" value=\"1\" checked>ON");
   client.println("<input type=\"radio\"
   name=\"switch\" value=\"0\"
   onclick=\"submit();\" >OFF");
}
else {
   client.println("<input type=\"radio\"
   name=\"switch\" value=\"1\"
   onclick=\"submit();\" >ON");
   client.println("<input type=\"radio\"
   name=\"switch\" value=\"0\" checked>OFF");
}
client.println("</form>");
```

Also, depending on the radio button selection, the browser will re-render the radio buttons using the server response to reflect the current status of the PowerSwitch Tail.

Adding a Cascade Style Sheet to the web user interface

Cascade Style Sheet (CSS) defines how HTML elements are to be displayed. Metro UI CSS (`https://metroui.org.ua/`) is a cascade style sheet that can be used to apply Windows 8-like style to your HTML elements.

The following Arduino sketch applies Windows 8-like style to the radio button group:

1. Open your Arduino IDE and paste the code from the sketch named `B04844_01_05.ino` from the code folder of this chapter.

2. Between the `<head></head>` tags we have first included the JQuery library which consists of a rich set of JavaScript functions:

```
client.println("<script
src=\"https://metroui.org.ua/js/jquery-
2.1.3.min.js\"></script>");
```

3. Then, we have included `metro.js` and `metro.css` from the `https://metroui.org.ua` website:

```
client.println("<script
src=\"https://metroui.org.ua/js/metro.js\"></script>");
client.println("<link rel=\"stylesheet\"
href=\"https://metroui.org.ua/css/metro.css\">");
```

Upload the sketch on your Arduino board and play with the new look and feel. You can modify the other HTML elements and even use the radio buttons by referring to the MetroUI CSS website documentation at `https://metroui.org.ua/`.

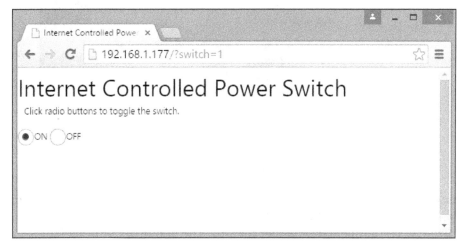

MetroUI CSS style applied to radio buttons

Finding the MAC address and obtaining a valid IP address

To work with this project, you must know your Arduino Ethernet Shield's MAC address and IP address to communicate properly over the Internet.

Finding the MAC address

Current Arduino Ethernet Shields come with a dedicated and uniquely assigned 48-bit MAC (Media Access Control) address which is printed on the sticker. Write down your Ethernet shield's MAC address so you can refer to it later. The following image shows an Ethernet shield with the MAC address of **90-A2-DA-0D-E2-CD**:

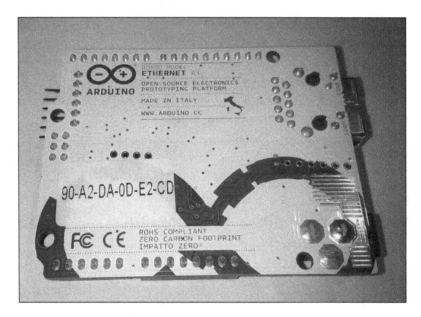

You can rewrite your Arduino Ethernet Shield's MAC address using hexadecimal notations, as in 0x90, 0xA2, 0xDA, 0x0D, 0xE2 and 0xCD, with the leading 0x notation recognized by C compilers (remember that the Arduino programming language is based on C) and assembly languages.

If not present, you can use one that does not conflict with your network.
For example:

```
byte mac[] = { 0xDE, 0xAD, 0xBE, 0xEF, 0xFE, 0xED };
```

Obtaining an IP address

You can assign an IP address to your Arduino Ethernet Shield by one of the following methods:

- Using the network router or switch to assign a static IP address to your Ethernet shield.

- Using DHCP (Dynamic Host Configuration Protocol) to dynamically assign an IP address to your Ethernet shield. In this chapter, we will only discuss how to assign an IP address using DHCP.

The network devices we will use for this experiment are the following:

- Huawei E517s-920 4G Wi-Fi Router

- DELL computer with Windows 8.1 installed and Wi-Fi connected

- Nokia Lumia phone with Windows 8.1 installed and Wi-Fi connected

- Arduino Ethernet Shield connected to the Wi-Fi router's LAN port using an Ethernet cable

Assigning a static IP address

The following steps will explain how to determine your network IP address range with a Windows 8.1 installed computer, and select a valid static IP address.

1. Open **Network and Sharing Center** in Control Panel:

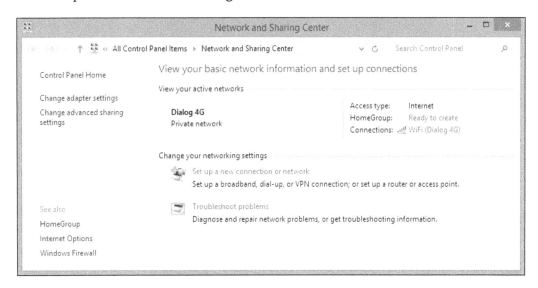

2. Click on **Connections**. The **Connection Status** dialog box will appear, as shown here:

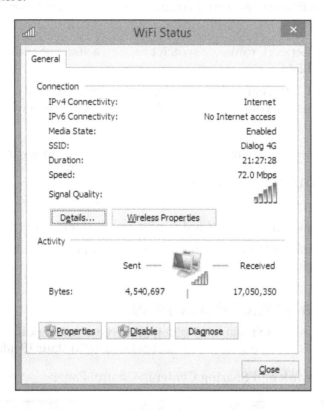

3. Click on the **Details...** button. The **Network Connection Details** dialog box will appear, as shown in the following screenshot:

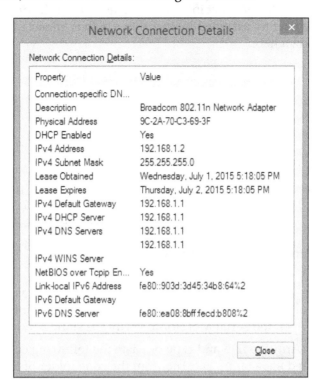

4. The IPv4 address assigned to the Windows 8.1 computer by the Wireless router is 192.168.1.2. The IPv4 subnet mask is 255.255.255.0. So, the IP address range should be 192.168.1.0 to 192.168.1.255.

5. The Wi-Fi network used in this example currently has two devices connected, that is, a Windows 8.1 computer, and a Windows phone. After logging in to the wireless router product information page, under the device list, all the IP addresses currently assigned by the router to the connected devices can be seen, as shown here:

Device List

Index	Computer Name	MAC Address	IP Address	Lease Time	Status	Type	Operation
1	DELL	9C:2A:70:C 3:69:3F	192.168.1. 2	0 days 22 hours 58 minutes 38 seconds	Active	Wi-Fi	Kick Out
2	Windows-Phone	A8:44:81:4 3:AD:C4	192.168.1. 3	0 days 22 hours 59 minutes 51 seconds	Active	Wi-Fi	Kick Out

6. Now, we can choose any address except 192.168.1.1, 192.168.1.2, and 192.168.1.3.

7. Let's assign 192.168.1.177 to the Arduino Ethernet Shield as a static IP address using the following sketch. Upload the following sketch into your Arduino board and open the Serial Monitor to verify the static IP address assigned.

8. Open your Arduino IDE and type or paste the following code from the sketch named B04844_01_06.ino from the code folder of this chapter.

```
#include <SPI.h>
#include <Ethernet.h>

byte mac[] = { 0xDE, 0xAD, 0xBE, 0xEF, 0xFE, 0xED };
byte ip[] = { 192, 168, 1, 177 };

EthernetServer server(80);

void setup()
{
  Serial.begin(9600);

  Ethernet.begin(mac, ip);
  server.begin();
  Serial.print("IP Address: ");
  Serial.println(Ethernet.localIP());

}

void loop () {}
```

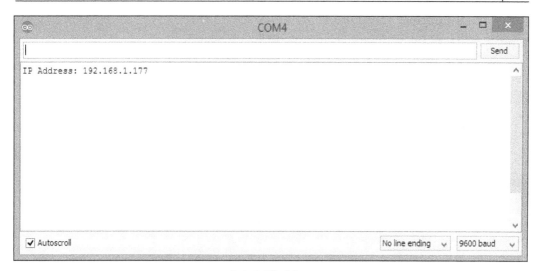

A static IP address

Obtaining an IP address using DHCP

The DHCP can be used to automatically assign a valid IP address to the Arduino Ethernet Shield. The only address you need is the MAC address of the Ethernet shield. Pass the MAC address as a parameter to the `Ethernet.begin()` method.

Upload the following Arduino sketch to your Arduino board, and open the Arduino Serial Monitor to see the auto-assigned IP address by the DHCP. Use this IP address to access your Ethernet shield through the Internet. Remember, this IP address may be changed at the next start up or reset.

Open your Arduino IDE and type or paste the following code from the sketch named `B04844_01_07.ino` from the code folder of this chapter:

```
#include <SPI.h>
#include <Ethernet.h>

byte mac[] = { 0xDE, 0xAD, 0xBE, 0xEF, 0xFE, 0xED };

EthernetServer server(80);

void setup()
{
  Serial.begin(9600);

  Ethernet.begin(mac);
  server.begin();
```

```
    Serial.print("IP Address: ");
    Serial.println(Ethernet.localIP());

}

void loop () {}
```

DHCP assigned IP address

Summary

In this chapter, you have gained a lot, and built your first Arduino **Internet of Things (IoT)** project, an internet controlled power switch, which is very smart. Using your creative knowledge, you can take this project to a more advanced level by adding many more functionalities, such as an LCD screen to the switch to display the current status and received user requests, or a feedback LED to show different statuses, and so on.

In the next chapter, you will learn how to build a Wi-Fi signal strength notification system using Arduino wearable and Internet of Things. Use the basic knowledge about Arduino IoT gained from this chapter to build the next project more successfully. Always be creative!

2

Wi-Fi Signal Strength Reader and Haptic Feedback

When designing an embedded system with Internet connectivity using Wi-Fi, reading the Wi-Fi connections receiving signal allows the user to determine the available Internet connectivity and signal strength. Most devices show the signal strength to the consumer using a simple bar graph or something similar. In this project, however, we look into how to notify the signal strength level using a different kind of mechanism to the user: the haptic feedback.

Another technique is to send the Wi-Fi signal strength level over the Internet, which allows you to measure signal strength even in unreachable locations. In the previous chapter, you learned about Arduino Ethernet Web server. Here, similar implementations will be used.

In this chapter, you will do the following:

- Learn about Arduino WiFi Shield basics and stacking with an Arduino UNO board
- Learn how to read the receiving radio signal strength level using RSSI
- Learn about vibration motors and haptic feedback
- Learn about haptic motor controllers and the Adafruit haptic library
- Write a simple web server to display the strength level of the received radio signal using a simple HTML web page

Prerequisites

To complete this project, you may require some open source hardware, software, tools, and good soldering skills. Let's dive in one step at a time.

- Arduino UNO board (http://store.arduino.cc/product/A000066)
- Arduino WiFi Shield (http://store.arduino.cc/product/A000058)
- Vibrating Mini Motor Disc (http://www.adafruit.com/product/1201)
- Adafruit DRV2605L Haptic Motor Controller (http://www.adafruit.com/product/2305)
- USB A to B cable
- Wall adapter power supply 9V DC 650mA

Arduino WiFi Shield

Arduino WiFi Shield allows you to connect your Arduino board to the Internet wirelessly. In the previous chapter, you learned how to connect the Arduino board to the Internet using an Ethernet shield with a wired connection. Unlike a wired connection, a wireless connection provides us with increased mobility within the Wi-Fi signal range, and the ability to connect to other Wi-Fi networks automatically, if the current network loses connection or has insufficient radio signal strength. Most of the mechanisms can be manipulated using the Arduino Wi-Fi library, a well-written piece of program sketch. The following image shows the top view of an Arduino WiFi Shield. Note that two rows of wire wrap headers are used to stack with the Arduino board.

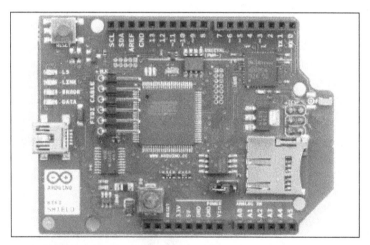

Arduino WiFi Shield (top view) Image courtesy of Arduino (https://www.arduino.cc) and license at http://creativecommons.org/licenses/by-sa/3.0/

The following image shows the bottom view of an Arduino WiFi Shield:

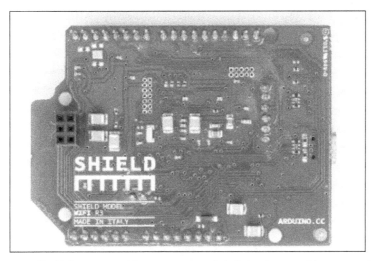

Arduino WiFi Shield (bottom view) Image courtesy of Arduino (https://www.arduino.cc)
and license at http://creativecommons.org/licenses/by-sa/3.0/

Firmware upgrading

Before using the Arduino WiFi Shield with this project, upgrade its firmware to version 1.1.0 or greater, as explained in the following official Arduino page at https://www.arduino.cc/en/Hacking/WiFiShieldFirmwareUpgrading.

The default factory-loaded firmware version 1.0.0 will not work properly with some of the Arduino sketches in this chapter.

Stacking the WiFi Shield with Arduino

Simply plug in to your Arduino WiFi Shield on top of the Arduino board using wire wrap headers so the pin layout of the Arduino board and the WiFi Shield will be exactly intact together.

Arduino WiFi Shield is stacked with Arduino UNO

Hacking an Arduino earlier than REV3

You can use the Arduino UNO REV3 board directly without any hacking for this project. However, you can still use an Arduino UNO board earlier than REV3 with a simple hack.

First, stack your Wi-Fi shield on the Arduino board, and then connect your Wi-Fi shield's IOREF pin to the 3.3V pin using a small piece of jumper wire.

The following image shows a wire connection from the **3.3V** pin to the **IOREF** pin.

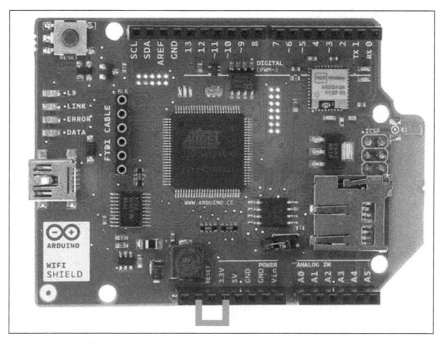

A jumper wire attached from 3.3V TO IOREF Image courtesy of Arduino (https://www.arduino.cc) and license at http://creativecommons.org/licenses/by-sa/3.0/

> **Warning!**
> Later, when you stack the hacked WiFi shield on an Arduino REV3 board, remember to remove the jumper wire. Otherwise, you will be shorting 3.3V to 5V through the **IOREF** pin.

Knowing more about connections

Your WiFi shield may have an SD card slot that communicates with your Arduino board via the digital pin 4. Arduino UNO communicates with the WiFi shield using digital pins 11, 12, and 13 over SPI bus. Also, the digital pin 10 is used as SS. Therefore, we will not use these pins with our project. However, you can use the digital pin 4 by using the following software hack.

```
pinMode(4, output);
digitalWrite(4, HIGH);
```

Fixing the Arduino WiFi library

Before getting started with the WiFi library, you have to apply the following fixes to some of the files inside the Arduino WiFi library:

1. Navigate to the `WiFi` folder in the libraries folder

2. Open the `wifi_drv.cpp` file located in the `utility` folder under `src`.

3. Find the `getCurrentRSSI()` function and modify it as follows:

```
int32_t WiFiDrv::getCurrentRSSI()
{
    startScanNetworks();
    WAIT_FOR_SLAVE_SELECT();
    // Send Command
    SpiDrv::sendCmd(GET_CURR_RSSI_CMD, PARAM_NUMS_1);

    uint8_t _dummy = DUMMY_DATA;
    SpiDrv::sendParam(&_dummy, 1, LAST_PARA);

    //Wait the reply elaboration
    SpiDrv::waitForSlaveReady();

    // Wait for reply
    uint8_t _dataLen = 0;
    int32_t rssi = 0;
    SpiDrv::waitResponseCmd(GET_CURR_RSSI_CMD, PARAM_NUMS_1,
    (uint8_t*)&rssi, &_dataLen);

    SpiDrv::spiSlaveDeselect();

    return rssi;
}
```

4. Save and close the file.

Connecting your Arduino to a Wi-Fi network

To connect your Arduino WiFi shield to a Wi-Fi network, you should have the SSID of any available Wi-Fi network. **SSID (Service Set Identifier)** is the name of the Wi-Fi network that you want to connect to your Arduino WiFi shield. Some Wi-Fi networks require a password to connect it with and some are not, which means open networks.

The Arduino WiFi library provides an easy way to connect your WiFi shield to a Wi-Fi network with the `WiFi.begin()` function. This function can be called in different ways depending on the Wi-Fi network that you want to connect to.

`WiFi.begin();` is only for initializing the Wi-Fi shield and called without any parameters.

1. `WiFi.begin(ssid);` connects your WiFi shield to an Open Network using only the `SSID` of the network, which is the name of the network. The following Arduino sketch will connect your Arduino WiFi shield to an open Wi-Fi network which is not password protected and anyone can connect. We assume that you have a Wi-Fi network configured as OPEN and named as `MyHomeWiFi`. Open a new Arduino IDE and copy the sketch named `B04844_02_01.ino` from the `Chapter 2` sample code folder.

```
#include <SPI.h>
#include <WiFi.h>

char ssid[] = "MyHomeWiFi";
int status = WL_IDLE_STATUS;

void setup(){
  Serial.begin(9600);
if (WiFi.status() == WL_NO_SHIELD){
  Serial.println("No WiFi shield found");   while(true);
}

while ( status != WL_CONNECTED){
  Serial.print("Attempting to connect to open SSID: ");
  Serial.println(ssid);
  status = WiFi.begin(ssid);

  delay(10000);
}

Serial.print("You're connected to the network");
}

void loop (){

}
```

2. Modify the following line of the code according to your Wi-Fi network's name.

```
char ssid[] = "MyHomeWiFi";
```

3. Now verify and upload the sketch in to your Arduino board and then open the Arduino Serial Monitor. The Arduino Serial Monitor will display the status about the connection at the time it was connected similar to follows.

Attempting to connect to open SSID: MyHomeWiFi

You're connected to the network

`WiFi.begin(ssid,pass);` connects your WiFi shield to a **WPA2 (Wi-Fi Protected Access II)** personal encrypted secured Wi-Fi network using SSID and password. The shield will not connect to Wi-Fi networks that are encrypted using WPA2 Enterprise Encryption. We assume that you have a Wi-Fi network configured as `WAP2` and named as `MyHomeWiFi`.

1. Open a new Arduino IDE and copy the sketch named `B04844_02_02.ino` from the `Chapter 2` sample code folder.

```
#include <SPI.h>
#include <WiFi.h>

char ssid[] = "MyHomeWiFi";
char pass[] = "secretPassword";
int status = WL_IDLE_STATUS;

void setup(){
Serial.begin(9600);
if (WiFi.status() == WL_NO_SHIELD) {
Serial.println("WiFi shield not present");
while(true);
   }
while ( status != WL_CONNECTED) {Serial.print("Attempting to
connect to WPA SSID: ");
Serial.println(ssid);

status = WiFi.begin(ssid, pass);

delay(10000);
   }

Serial.print("You're connected to the network");

}
```

```
void loop(){

}
```

2. Modify the following line of the code according to your Wi-Fi network's name.

```
char ssid[] = "MyHomeWiFi";
```

3. Now verify and upload the sketch in to your Arduino board and then open the Arduino Serial Monitor. The Arduino Serial Monitor will display the status about the connection at the time it was connected similar to follows.

Attempting to connect to WPA SSID: MyHomeWiFi

You're connected to the network

`WiFi.begin(ssid, keyIndex, key);` is only for use with WEP encrypted Wi-Fi networks. WEP networks can have up to four passwords in hexadecimals that are known as keys. Each key is assigned a `Key Index` value. Configure your Wi-Fi network as a WEP encryption and upload the following sketch into your Arduino board. But remember the WEP is not secure at all and don't use it with your Wi-Fi networks. Instead of that use WPA2 encryption which is highly recommended.

1. We assume that you have a Wi-Fi network configured as `WPE` and named as `MyHomeWiFi`. Change the configuration back to the WPA2 as quickly as possible after testing the following code snippet. Open a new Arduino IDE and copy the sketch named `B04844_02_03.ino` from the `Chapter 2` sample code folder.

```
#include <SPI.h>
#include <WiFi.h>

char ssid[] = "MyHomeWiFi";
char key[] = "D0D0DEADF00DABBADEAFBEADED";
int keyIndex = 0;
int status = WL_IDLE_STATUS;

void setup(){

  Serial.begin(9600);
  if (WiFi.status() == WL_NO_SHIELD)  {
    Serial.println("WiFi shield not present");        while(true);
  }
  while ( status != WL_CONNECTED)  {    Serial.print("Attempting
to connect to WEP network, SSID: ");
```

```
        Serial.println(ssid);
        status = WiFi.begin(ssid, keyIndex, key);
            delay(10000);
    }

    Serial.print("You're connected to the network");
    }


void loop(){

}
```

2. Modify the following line of the code according to your Wi-Fi network's name:
   ```
   char ssid[] = "MyHomeWiFi";
   ```

3. Now verify and upload the sketch in to your Arduino board and then open the Arduino Serial Monitor. The Arduino Serial Monitor will display the status about the connection at the time it was connected similar to follows.

   Attempting to connect to WEP network, SSID: MyHomeWiFi

   You're connected to the network

Wi-Fi signal strength and RSSI

The Arduino WiFi library provides us with a simple way to get the Wi-Fi signal strength in decibels ranging from 0 to -100 (minus 100). You can use the `WiFi.RSSI()` function to get the radio signal strength of the currently connected network or any specified network. You can read more about **Received Signal Strength Indication (RSSI)** at `https://en.wikipedia.org/wiki/Received_signal_strength_indication`.

The `WiFi.RSSI()` function can be called with following parameters:

- `WiFi.RSSI();`:This will return the signal strength of the currently connected Wi-Fi network.

- `WiFi.RSSI(WiFi Access Point);`: This will return the signal strength of a specified Wi-Fi network. Wi-Fi Access Point is the name of the Wi-Fi network. For example, `MyHomeWiFi`.

Reading the Wi-Fi signal strength

Now we will write an Arduino sketch to get the RSSI value of the currently connected Wi-Fi network.

1. Open a new Arduino IDE and copy the sketch named `B04844_02_04.ino` from the `Chapter 2` sample code folder.

```
#include <SPI.h>
#include <WiFi.h>

char ssid[] = "MyHomeWiFi";
char pass[] = "secretPassword";

void setup()
{
  WiFi.begin(ssid, pass);
    if (WiFi.status() != WL_CONNECTED)    {     Serial.
println("Couldn't get a wifi connection");
      while(true);
    }        else
        {
      long rssi = WiFi.RSSI();
      Serial.print("RSSI: ");
      Serial.print(rssi);
      Serial.println(" dBm");
    }
  }
}

void loop (){
}
```

2. Modify the following line of the code according to your Wi-Fi network's name.

```
char ssid[] = "MyHomeWiFi";
```

3. Now verify and upload the sketch in to your Arduino board and then open the Arduino Serial Monitor.

4. The Arduino Serial Monitor will display the received signal in dBm (decibel-milliwatts) for the currently connected Wi-Fi network similar to the following:

However, note that this will only provide the signal strength at the moment the WiFi shield was connected to the Wi-Fi network.

In the next Arduino sketch, we are going to look at how to display the Wi-Fi signal strength and update it periodically.

1. Open a new Arduino IDE and copy the sketch named B04844_02_05.ino from the Chapter 2 sample code folder.

```
#include <SPI.h>
#include <WiFi.h>

char ssid[] = "MyHomeWiFi";
char pass[] = "secretPassword";

void setup()
{
WiFi.begin(ssid, pass);
}

void loop (){
```

```
    if (WiFi.status() != WL_CONNECTED)    {    Serial.
println("Couldn't get a wifi connection");
    while(true);
    }
  else
    {
    long rssi = WiFi.RSSI();
    Serial.print("RSSI: ");
    Serial.print(rssi);
    Serial.println(" dBm");
    }

delay(10000);//waits 10 seconds and update

}
```

2. Modify the following line of the code according to your WiFi network's name:

```
char ssid[] = "MyHomeWiFi";
```

3. Now verify and upload the sketch in to your Arduino board and then open the Arduino Serial Monitor.

4. The Arduino Serial Monitor will display the received signal in dBm (decibel-milliwatts) for the currently connected Wi-Fi network similar to the following:

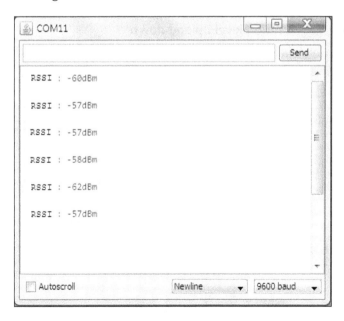

In the next section of this chapter, we will look at how to integrate a vibrator to the Arduino WiFi shield and output advanced vibration patterns according to the current RSSI value.

Haptic feedback and haptic motors

Haptic feedback is the way to convey information to users using advanced vibration patterns and waveforms. Earlier consumer electronic devices communicated with their users using audible and visual alerts, but now things have been replaced with vibrating alerts through haptic feedback.

In a haptic feedback system, the vibrating component can be a vibration motor or a linear resonant actuator. The motor is driven by a special hardware called the haptic controller or haptic driver. Throughout this chapter we use the term **vibrator** for the vibration motor.

Getting started with the Adafruit DRV2605 haptic controller

Adafruit DRV2605 haptic controller is an especially designed motor controller for controlling haptic motors. With a haptic controller, you can make various effects using a haptic motor such as:

- Ramping the vibration level up and down
- Click, double-click, and triple-click effects
- Pulsing effects
- Different buzzer levels
- Vibration following a musical/audio input

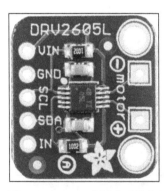

The DRV2605 breakout board (top view) Image courtesy of Adafruit Industries (https://www.adafruit.com)

Selecting a correct vibrator

Vibrators come with various shapes and driving mechanisms. Some of them support haptic feedback while some do not. Before purchasing a vibrator, check the product details carefully to determine whether it supports haptic feedback. For this project, we will be using a simple vibrating mini motor disc, which is a small disc-shaped motor. It has negative and positive leads to connect with the microcontroller board.

The following image shows a vibrator with positive and negative wires soldered:

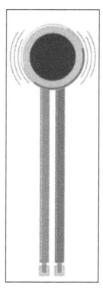

Fritzing representation of a vibrator

Connecting a haptic controller to Arduino WiFi Shield

Use the following steps to connect the DRV2605 haptic controller to Arduino WiFi Shield:

1. Solder headers to the DRV2605 breakout board, connect it to a breadboard and then user jumper wires for the connection to the Arduino.

2. Connect the **VIN** pin of the DRV2605 breakout board to the **5V** pin of the Arduino WiFi Shield.

3. Connect the **GND** pin of the DRV2605 breakout to the **GND** pin of the Arduino WiFi Shield.

4. Connect the **SCL** pin of the DRV2605 breakout board to the Analog 5 (**A5**) pin of the Arduino WiFi Shield.

5. Finally, connect the **SDA** pin of the DRV2605 breakout board to the Analog 4 (**A4**) pin of the Arduino WiFi Shield.

The following image shows the connection between DRV2605 breakout board and Arduino WiFi shield:

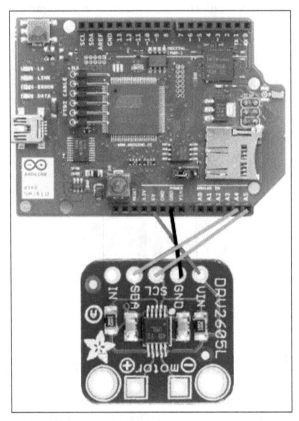

Image courtesy of Arduino (https://www.arduino.cc) and license at http://creativecommons.org/ licenses/by-sa/3.0/, and Adafruit Industries (https://www.adafruit.com)

Soldering a vibrator to the haptic controller breakout board

On the DRV2605 breakout board, you can see two square shaped soldering pads marked as + and - along with the **motor** text label. This is the place where we are going to solder the vibrator. Generally vibrators have two presoldered wires, red and black.

- Solder the red wire of the vibrator to the + soldering pad of the breakout board
- Solder the blue wire of the vibrator to the − soldering pad of the breakout board

The following image shows the final connection between DRV2605 breakout board, Arduino WiFi shield and the vibrator:

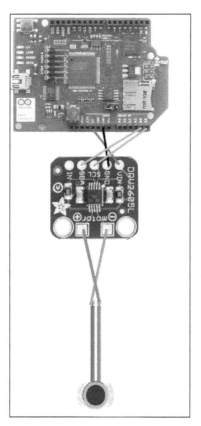

Image courtesy of Arduino (https://www.arduino.cc) and license at http://creativecommons.org/licenses/by-sa/3.0/, and Adafruit Industries (https://www.adafruit.com)

Downloading the Adafruit DRV2605 library

You can download the latest version of the Adafruit DRV2605 library from the GitHub repository by navigating to the following URL: `https://github.com/adafruit/Adafruit_DRV2605_Library`.

The Adafruit DRV2605 library at GitHub

After navigating to the earlier URL, follow these steps to download Adafruit DRV2605 library:

1. Click on the **Download Zip** button.

2. After downloading the ZIP file, extract it to your local drive and rename it as `Adafruit_DRV2605`. Then copy or move the folder inside the Arduino libraries folder. Finally, restart the Arduino IDE.

3. Open the sample sketch included with the library by clicking on **File | Examples | Adafruit_DRV2605 | basic** and upload it to your Arduino board. The sketch will play 116 vibration effects defined in the DRV2605 library from effect number 1 to 116 in order.

You can download the datasheet for DRV2605 Haptic Driver from `http://www.ti.com/lit/ds/symlink/drv2605.pdf` and refer to pages 55-56 for the full set of 123 vibration effects. The DRV2605 haptic driver is manufacturing by Texas Instruments.

Making vibration effects for RSSI

Now, we will learn how to make different vibration effects depending on the **Received Signal Strength Indication (RSSI)**. Typically, RSSI value rages from 0 to -100. The higher the value, the stronger the signal reception where 0 is the highest value. Therefore, we can logically check the RSSI value return by the `WiFi.RSSI()` function and play vibration effects accordingly.

In the following example, we will play the first 10 vibration effects according to the RSSI value output by the Arduino WiFi shield. See the following chart for the RSSI value range for each vibration effect:

Effect Number	RSSI	
1	0	-10
2	-11	-20
3	-21	-30
4	-31	-40
5	-41	-50
6	-51	-60
7	-61	-70
8	-71	-80
9	-81	-90
10	-91	-100

Following steps shows how to generate different vibration effects according to the RSSI strength of the currently connected Wi-Fi network.

1. Open a new Arduino IDE and copy the sketch named `B04844_02_06.ino` from the `Chapter 2` sample code folder.

2. Modify the following line of the code according to your Wi-Fi network's name:

   ```
   char ssid[] = "MyHomeWiFi";
   ```

3. Following line maps RSSI output to the value range from 1 to 10 using the `map()` function:

   ```
   int range = map(rssi, -100, 0, 1, 10);
   ```

4. Set the vibration effect using the `setWaveform(slot, effect)` function by passing the parameters such as slot number and effect number. Slot number starts from 0 and effect number can be found in the waveform library effect list.

5. Finally call the `go()` function to play the effect.

    The following code block shows first how to set and play the waveform `double click - 100%`:

    ```
    drv.setWaveform(0, 10);  // play double click - 100%drv.
    setWaveform(1, 0);    // end waveform
    drv.go(); // play the effect!
    ```

6. Verify and upload the sketch in to your Arduino board. Now touch the vibrator and feel the different vibration effects according to the variations of WiFi signal strength of the currently connected network. You can test this by moving your Wi-Fi router away from the Arduino WiFi shield.

Downloading the example code

You can download the example code files from your account at `http://www.packtpub.com` for all the Packt Publishing books you have purchased. If you purchased this book elsewhere, you can visit `http://www.packtpub.com/support` and register to have the files e-mailed directly to you.

Implementing a simple web server

The Arduino WiFi Shield can also be configured and programmed as a web server to serve client requests similar to the Arduino Ethernet shield. In the next step, we will be making a simple web server to send Wi-Fi signal strength over the Internet to a client. This requires that the WiFi shield has 1.1.0 firmware to work. The default factory loaded version 1.0.0 will not work. (See the *Firmware upgrading* section.)

Reading the signal strength over Wi-Fi

To read the signal strength over Wi-Fi:

1. Open a new Arduino IDE and copy the sketch named `B04844_02_07.ino` from the `Chapter 2` sample code folder.

2. Verify and upload the Arduino sketch in to the Arduino board. Type the IP address of your WiFi shield in your web browser and hit the *Enter* key. The web page will load and display the current RSSI of the Wi-Fi network and refresh every 20 seconds. If you don't know the IP address of your Arduino WiFi shield assigned by the DHCP, open the Arduino Serial Monitor and you can find it from there.

Summary

In this chapter, we learnt how to read Wi-Fi signal strength with a WiFi shield and make haptic feedback using a vibration motor according to the Wi-Fi signal strength. Further, we learned to use haptic feedback libraries to make feedback patterns.

In the next chapter, we will learn how to select and use a water flow sensor, and then connect it with Arduino and calibrate it, and also how to calculate and display the values on a LCD screen and store data in the cloud.

3

Internet-Connected Smart Water Meter

For many years and even now, water meter readings have been collected manually. To do this, a person has to visit the location where the water meter is installed. In this chapter, you will learn how to make a smart water meter with an LCD screen that has the ability to connect to the internet and serve meter readings to the consumer through the Internet.

In this chapter, you shall do the following:

- Learn about water flow sensors and its basic operation
- Learn how to mount and plumb a water flow meter on and into the pipeline
- Read and count the water flow sensor pulses
- Calculate the water flow rate and volume
- Learn about LCD displays and connecting with Arduino
- Convert a water flow meter to a simple web server and serve meter readings through the Internet

Prerequisites

- An Arduino UNO R3 board (http://store.arduino.cc/product/A000066)
- Arduino Ethernet Shield R3 (https://www.adafruit.com/products/201)
- A liquid flow sensor (http://www.futurlec.com/FLOW25L0.shtml)
- A Hitachi HD44780 DRIVER compatible LCD Screen (16 x 2) (https://www.sparkfun.com/products/709)

- A 10K ohm resistor
- A 10K ohm potentiometer (`https://www.sparkfun.com/products/9806`)
- Few Jumper wires with male and female headers (`https://www.sparkfun.com/products/9140`)
- A breadboard (`https://www.sparkfun.com/products/12002`)

Water flow sensors

The heart of a water flow sensor consists of a Hall effect sensor (`https://en.wikipedia.org/wiki/Hall_effect_sensor`) that outputs pulses for magnetic field changes. Inside the housing, there is a small pinwheel with a permanent magnet attached to it. When the water flows through the housing, the pinwheel begins to spin, and the magnet attached to it passes very close to the Hall effect sensor in every cycle. The Hall effect sensor is covered with a separate plastic housing to protect it from the water. The result generates an electric pulse that transitions from low voltage to high voltage, or high voltage to low voltage, depending on the attached permanent magnet's polarity. The resulting pulse can be read and counted using the Arduino.

For this project, we will use a Liquid Flow sensor from *Futurlec* (`http://www.futurlec.com/FLOW25L0.shtml`). The following image shows the external view of a Liquid Flow Sensor:

Liquid flow sensor – the flow direction is marked with an arrow

The following image shows the inside view of the liquid flow sensor. You can see a pinwheel that is located inside the housing.

Pinwheel attached inside the water flow sensor

Wiring the water flow sensor with Arduino

The water flow sensor that we are using with this project has three wires, which are the following:

- Red (or it may be a different color) wire, which indicates the Positive terminal
- Black (or it may be a different color) wire, which indicates the Negative terminal
- Brown (or it may be a different color) wire, which indicates the DATA terminal

All three wire ends are connected to a JST connector. Always refer to the datasheet of the product for wiring specifications before connecting them with the microcontroller and the power source.

When you use jumper wires with male and female headers, do the following:

1. Connect positive terminal of the water flow sensor to Arduino **5V**.
2. Connect negative terminal of the water flow sensor to Arduino **GND**.
3. Connect DATA terminal of the water flow sensor to Arduino digital pin **2**.

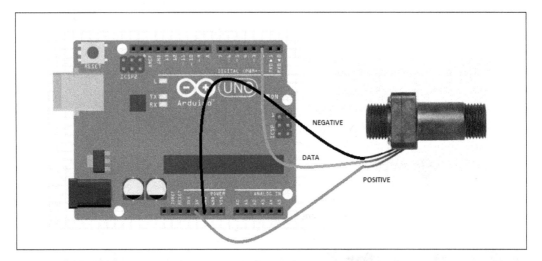

Water flow sensor connected with Arduino Ethernet Shield using three wires

You can directly power the water flow sensor using Arduino since most residential type water flow sensors operate under 5V and consume a very low amount of current. Read the product manual for more information about the supply voltage and supply current range to save your Arduino from high current consumption by the water flow sensor. If your water flow sensor requires a supply current of more than 200mA or a supply voltage of more than 5v to function correctly, then use a separate power source with it.

The following image illustrates jumper wires with male and female headers:

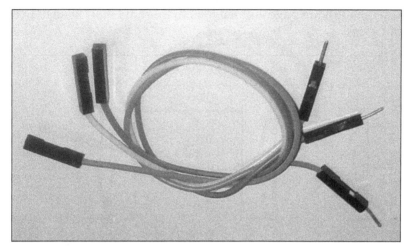

Jumper wires with male and female headers

Reading pulses

The water flow sensor produces and outputs digital pulses that denote the amount of water flowing through it. These pulses can be detected and counted using the Arduino board.

Let's assume the water flow sensor that we are using for this project will generate approximately 450 pulses per liter (most probably, this value can be found in the product datasheet). So 1 pulse approximately equals to [1000 ml/450 pulses] 2.22 ml. These values can be different depending on the speed of the water flow and the mounting polarity of the water flow sensor.

Arduino can read digital pulses generating by the water flow sensor through the DATA line.

Rising edge and falling edge

There are two type of pulses, as listed here:.

- **Positive-going pulse**: In an idle state, the logic level is normally LOW. It goes HIGH state, stays there for some time, and comes back to the LOW state.

- **Negative-going pulse**: In an idle state, the logic level is normally HIGH. It goes LOW state, stays LOW state for time, and comes back to the HIGH state.

The rising and falling edges of a pulse are vertical. The transition from LOW state to HIGH state is called **rising edge** and the transition from HIGH state to LOW state is called **falling edge**.

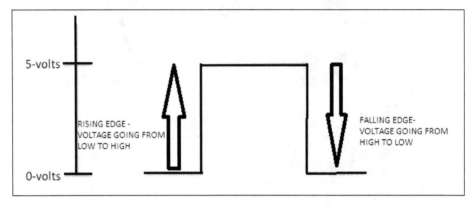

Representation of Rising edge and Falling edge in digital signal

You can capture digital pulses using either the rising edge or the falling edge. In this project, we will use the rising edge.

Reading and counting pulses with Arduino

In the previous step, you attached the water flow sensor to Arduino UNO. The generated pulse can be read by Arduino digital pin 2 and the interrupt 0 is attached to it.

The following Arduino sketch will count the number of pulses per second and display it on the Arduino Serial Monitor:

1. Open a new Arduino IDE and copy the sketch named `B04844_03_01.ino` from the `Chapter 3` sample code folder.

2. Change the following pin number assignment if you have attached your water flow sensor to a different Arduino pin:

```
int pin = 2;
```

3. Verify and upload the sketch on the Arduino board:

```
int pin = 2; //Water flow sensor attached to digital pin 2
volatile unsigned int pulse;
const int pulses_per_litre = 450;

void setup()
{
    Serial.begin(9600);

    pinMode(pin, INPUT);
    attachInterrupt(0, count_pulse, RISING);
}

void loop()
{
    pulse = 0;
    interrupts();
    delay(1000);
    noInterrupts();

    Serial.print("Pulses per second: ");
    Serial.println(pulse);
}

void count_pulse()
{
    pulse++;
}
```

4. Open the Arduino Serial Monitor and blow air through the water flow sensor using your mouth.

5. The number of pulses per second will print on the Arduino Serial Monitor for each loop, as shown in the following screenshot:

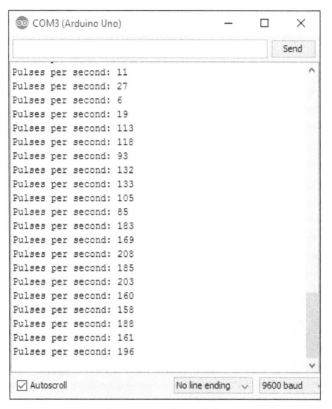

Pulses per second in each loop

The `attachInterrupt()` function is responsible for handling the `count_pulse()` function. When the `interrupts()` function is called, the `count_pulse()` function will start to collect the pulses generated by the liquid flow sensor. This will continue for 1000 milliseconds, and then the `noInterrupts()` function is called to stop the operation of `count_pulse()` function. Then, the pulse count is assigned to the pulse variable and prints it on the serial monitor. This will repeat again and again inside the loop() function until you press the reset button or disconnect the Arduino from the power.

Calculating the water flow rate

The water flow rate is the amount of water flowing in at a given point of time and can be expressed in gallons per second or liters per second. The number of pulses generated per liter of water flowing through the sensor can be found in the water flow sensor's specification sheet. Let's say there are m pulses per liter of water.

You can also count the number of pulses generated by the sensor per second: Let's say there are n pulses per second.

The water flow rate R can be expressed as:

$$R = \frac{n \text{ (pulse per second)}}{m \text{ (pulse per litre)}}$$

In liters per second

Also, you can calculate the water flow rate in liters per minute using the following formula:

$$R = \frac{n * 60 \text{ (pulse per minute)}}{m \text{ (pulse per litre)}}$$

For example, if your water flow sensor generates 450 pulses for one liter of water flowing through it, and you get 10 pulses for the first second, then the elapsed water flow rate is:

10/450 = 0.022 liters per second or *0.022 * 1000 = 22* milliliters per second.

The following steps will explain you how to calculate the water flow rate using a simple Arduino sketch:

1. Open a new Arduino IDE and copy the sketch named `B04844_03_02.ino` from the `Chapter 3` sample code folder.

2. Verify and upload the sketch on the Arduino board.

3. The following code block will calculate the water flow rate in milliliters per second:

```
Serial.print("Water flow rate: ");
Serial.print(pulse * 1000/pulses_per_litre);
Serial.println("milliliters per second");
```

4. Open the Arduino Serial Monitor and blow air through the water flow sensor using your mouth.

5. The number of pulses per second and the water flow rate in milliliters per second will print on the Arduino Serial Monitor for each loop, as shown in the following screenshot:

Pulses per second and water flow rate in each loop

Calculating the water flow volume

The water flow volume can be calculated by summing up the product of flow rate and the time interval:

$$Volume = \sum Flow\ Rate * Time\_Interval$$

The following Arduino sketch will calculate and output the total water volume since the device startup:

1. Open a new Arduino IDE and copy the sketch named `B04844_03_03.ino` from the `Chapter 3` sample code folder.

2. The water flow volume can be calculated using following code block:

```
volume = volume + flow_rate * 0.1; //Time Interval is 0.1 second

Serial.print("Volume: ");
Serial.print(volume);
Serial.println(" milliliters");
```

3. Verify and upload the sketch on the Arduino board.

4. Open the Arduino Serial Monitor and blow air through the water flow sensor using your mouth.

5. The number of pulses per second, water flow rate in milliliters per second, and total volume of water in milliliters will be printed on the Arduino Serial Monitor for each loop, as shown in the following screenshot:

Pulses per second, water flow rate and in each loop and sum of volume

To accurately measure water flow rate and volume, the water flow sensor needs to be carefully calibrated. The hall effect sensor inside the housing is not a precision sensor, and the pulse rate does vary a bit depending on the flow rate, fluid pressure, and sensor orientation. This topic is beyond the scope of this book.

Adding an LCD screen to the water meter

You can add an LCD screen to your newly built water meter to display readings, rather than displaying them on the Arduino serial monitor. You can then disconnect your water meter from the computer after uploading the sketch on to your Arduino.

Using a Hitachi HD44780 driver compatible LCD screen and Arduino Liquid Crystal library, you can easily integrate it with your water meter. Typically, this type of LCD screen has 16 interface connectors. The display has two rows and 16 columns, so each row can display up to 16 characters.

The following image represents the top view of a Hitachi HD44760 driver compatible LCD screen. Note that the 16-pin header is soldered to the PCB to easily connect it with a breadboard.

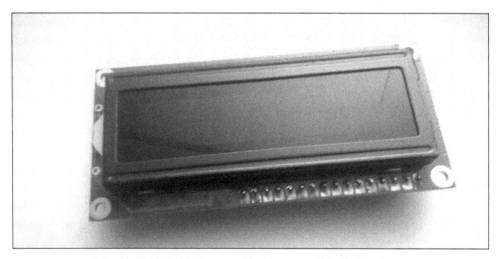

Hitachi HD44780 driver compatible LCD screen (16 x 2) — Top View

The following image represents the bottom view of the LCD screen. Again, you can see the soldered 16-pin header.

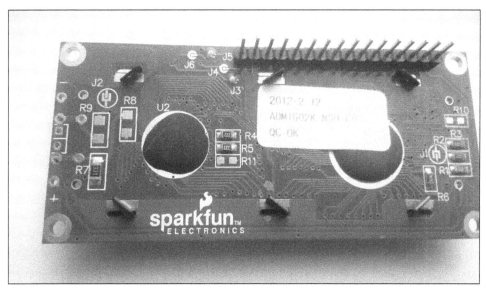

Hitachi HD44780 driver compatible LCD screen (16x2) – Bottom View

Wire your LCD screen with Arduino as shown in the next diagram. Use the 10k potentiometer to control the contrast of the LCD screen. Now, perform the following steps to connect your LCD screen with your Arduino:

1. LCD RS pin (pin number 4 from left) to Arduino digital pin 8.
2. LCD ENABLE pin (pin number 6 from left) to Arduino digital pin 7.
3. LCD READ/WRITE pin (pin number 5 from left) to Arduino GND.
4. LCD DB4 pin (pin number 11 from left) to Arduino digital pin 6.
5. LCD DB5 pin (pin number 12 from left) to Arduino digital pin 5.
6. LCD DB6 pin (pin number 13 from left) to Arduino digital pin 4.
7. LCD DB7 pin (pin number 14 from left) to Arduino digital pin 3.
8. Wire a 10K pot between Arduino +5V and GND, and wire its wiper (center pin) to LCD screen V0 pin (pin number 3 from left).
9. LCD GND pin (pin number 1 from left) to Arduino GND.
10. LCD +5V pin (pin number 2 from left) to Arduino 5V pin.

11. LCD Backlight Power pin (pin number 15 from left) to Arduino 5V pin.

12. LCD Backlight GND pin (pin number 16 from left) to Arduino GND.

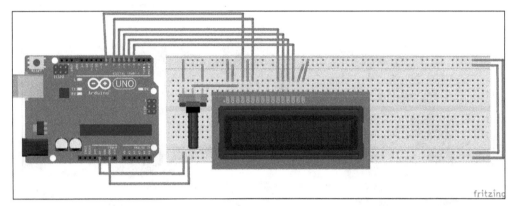

Fritzing representation of the circuit

13. Open a new Arduino IDE and copy the sketch named `B04844_03_04.ino` from the `Chapter 3` sample code folder.

14. First initialize the Liquid Crystal library using following line:

    ```
    #include <LiquidCrystal.h>
    ```

15. To create a new LCD object with following parameters, the syntax is `LiquidCrystal lcd (RS, ENABLE, DB4, DB5, DB6, DB7)`:

    ```
    LiquidCrystal lcd(8, 7, 6, 5, 4, 3);
    ```

16. Then initialize number of rows and columns in the LCD. Syntax is `lcd.begin(number_of_columns, number_of_rows)`:

    ```
    lcd.begin(16, 2);
    ```

17. You can set the starting location to print a text on the LCD screen using following function, syntax is `lcd.setCursor(column, row)`:

    ```
    lcd.setCursor(7, 1);
    ```

 Note that the column and row numbers are 0 index based and the following line will start to print a text in the intersection of the 8th column and 2nd row.

18. Then, use the `lcd.print()` function to print some text on the LCD screen:

    ```
    lcd.print(" ml/s");
    ```

19. Verify and upload the sketch on the Arduino board.

20. Blow some air through the water flow sensor using your mouth.

You can see some information on the LCD screen such as pulses per second, water flow rate, and total water volume from the beginning of the time:

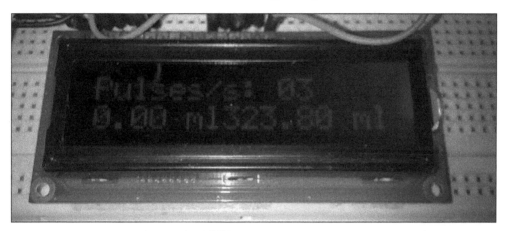

LCD screen output

Converting your water meter to a web server

In the previous steps, you learned how to display your water flow sensor's readings and calculate water flow rate and total volume on the Arduino serial monitor. In this step, you will learn how to integrate a simple web server to your water flow sensor and remotely read your water flow sensor's readings.

You can make an Arduino web server with Arduino WiFi Shield or Arduino Ethernet shield. The following steps will explain how to convert the Arduino water flow meter to a web server with Arduino Wi-Fi shield:

1. Remove all the wires you have connected to your Arduino in the previous sections in this chapter.

2. Stack the Arduino WiFi shield on the Arduino board using wire wrap headers. Make sure the Arduino WiFi shield is properly seated on the Arduino board.

3. Now, reconnect the wires from water flow sensor to the Wi-Fi shield. Use the same pin numbers as used in the previous steps.

4. Connect the 9VDC power supply to the Arduino board.

5. Connect your Arduino to your PC using the USB cable and upload the next sketch. Once the upload is completed, remove your USB cable from the Arduino.

6. Open a new Arduino IDE and copy the sketch named `B04844_03_05.ino` from the `Chapter 3` sample code folder.

7. Change the following two lines according to your WiFi network settings, as shown here:

```
char ssid[] = "MyHomeWiFi";
char pass[] = "secretPassword";
```

8. Verify and upload the sketch on the Arduino board.

9. Blow the air through the water flow sensor using your mouth, or it would be better if you can connect the water flow sensor to a water pipeline to see the actual operation with the water.

10. Open your web browser, type the WiFi shield's IP address assigned by your network, and hit the *Enter* key:

```
http://192.168.1.177
```

11. You can see your water flow sensor's pulses per second, flow rate, and total volume on the Web page. The page refreshes every 5 seconds to display updated information.

12. You can add an LCD screen to the Arduino WiFi shield as discussed in the previous step. However, remember that you can't use some of the pins in the Wi-Fi shield because they are reserved for SD (pin 4), SS (pin 10), and SPI (pin 11, 12, 13). We have not included the circuit and source code here in order to make the Arduino sketch simple.

A little bit about plumbing

Typically, the direction of the water flow is indicated by an arrow mark on top of the water flow meter's enclosure. Also, you can mount the water flow meter either horizontally or vertically according to its specifications. Some water flow meters can mount both horizontally and vertically.

You can install your water flow meter to a half-inch pipeline using normal BSP pipe connectors. The outer diameter of the connector is 0.78" and the inner thread size is half-inch.

The water flow meter has threaded ends on both sides. Connect the threaded side of the PVC connectors to both ends of the water flow meter. Use a thread seal tape to seal the connection, and then connect the other ends to an existing half-inch pipeline using PVC pipe glue or solvent cement.

Make sure that you connect the water flow meter with the pipe line in the correct direction. See the arrow mark on top of the water flow meter for flow direction.

BNC pipe line connector made by PVC

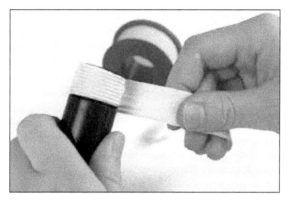

Securing the connection between the water flow meter and BNC pipe connector using thread seal

PVC solvent cement. Image taken from https://www.flickr.com/photos/ttrimm/7355734996/

Summary

In this chapter, you gained hands-on experience and knowledge about water flow sensors and counting pulses while calculating and displaying them. Finally, you made a simple web server to allow users to read the water meter through the Internet. You can apply this to any type of liquid, but make sure to select the correct flow sensor because some liquids react chemically with the material that the sensor is made of. You can Google and find which flow sensors support your preferred liquid type.

The next chapter will help you to make your own security camera with motion detection based on Arduino and Ethernet shield. You will be monitoring your home surroundings remotely in no time.

4
Arduino Security Camera with Motion Detection

Security is a concern for everyone. If you want to capture and record any activity within your home or office for security purposes, thousands of security camera models are available to fulfill the task. You can, however, make your own security camera, complete with Internet feedback and motion detection, and you can also access the camera images from your mobile's browser from anywhere in the world.

In this chapter, you will learn the following:

- How to use **TTL (Through The Lens)** Serial Camera directly with NTSC video screen. You can read more about TTL at `https://en.wikipedia.org/wiki/Through-the-lens_metering`.
- How to connect TTL Serial Camera to Arduino and Ethernet Shield.
- How to capture images with TTL Serial Camera.
- How to create Flickr and Temboo accounts and configure with Arduino Ethernet Shield.
- How to upload images to the Flickr using the Temboo cloud service.
- How to capture images with built-in motion sensor and upload them to the Flickr.

Prerequisites

The following materials will be needed to get started with the chapter:

- Arduino UNO Rev3 board (`https://store.arduino.cc/product/A000066`).

- Arduino Ethernet Shield Rev3 (`https://store.arduino.cc/product/A000072`).

- Arduino Ethernet board (optional) — if you use an Arduino Ethernet board, you do not need an Arduino UNO board. Arduino Ethernet board is a compact collection of an Arduino board and Ethernet Shield. (`https://www.sparkfun.com/products/11229`).

- Micro SD Card — Use 4GB Class 4 SDHC. (`https://www.adafruit.com/products/102`).

- TTL Serial Camera (`http://www.adafruit.com/products/397`).

- 9V DC power supply (`https://www.sparkfun.com/products/10273`).

- USB A-to-B cable (`https://www.sparkfun.com/products/512`).

- RCA jack (`https://www.sparkfun.com/products/8631`).

- Ethernet Cable (`https://www.sparkfun.com/products/8915`).

- Optional - NTSC/PAL TFT display – 4.3" Diagonal (`http://www.adafruit.com/product/946`).

- Jumper wires.

- Wire Strippers (`https://www.sparkfun.com/products/12630`).

- Soldering iron (EU:230V AC: `https://www.sparkfun.com/products/11650`, US 110V AC: `https://www.sparkfun.com/products/9507`).

Getting started with TTL Serial Camera

The heart of the TTL Serial Camera module (the product page at Adafruit named it as TTL Serial JPEG Camera) is the VIMICRO VC0706 Digital video Processor. The following are some of the features that a VC0706 digital video processor has:

- CMOS sensor interface and digital video input interface, so it can capture video using the CMOS sensor or external TV decoder

- Embedded TV encoder and video DAC, so it can directly output NTSC/PAL video streams to TV monitors and other 75 ohm display devices

- Preimage processing and M-JPEG compression ability
- NTSC video output resolution up to 712 x 486
- PAL video output resolution up to 704 x 576
- Maximum frame rate; 60fps @ 27MHz in NTSC and 50fps @ 27MHz in PAL
- Ability to change the brightness, saturation, and hue of images
- Auto brightness and auto contrast adjustment
- Motion detection

The VC0706 chipset specification mentioned that it supports both NTSC and PAL but the TTL Serial Camera module only implemented NTSC.

The TTL Serial Camera is only just a breakout board and has no wires, so you need to solder wires into the connection pads. It has five connection pads, which are 2mm apart from each other.

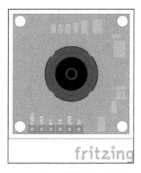

A Fritzing representation of TTL Serial Camera — top view

The pin labels and their core operation is listed as follows:

- **CVBS**: Outputs NTSC monochrome video stream
- **GND**: This is the NTSC video ground, located next to the CVBS pad
- **TX**: Data transmits from the module
- **RX**: Data reception to the module
- **GND**: Negative
- **+5V**: Positive

Wiring the TTL Serial Camera for image capturing

You need four wires if you are only planning to capture color images with a TTL serial camera.

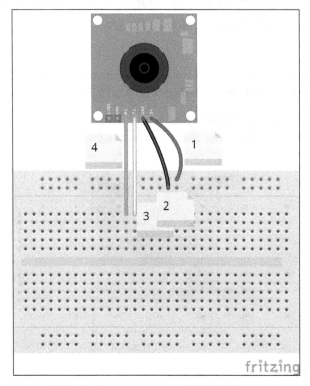

Wiring for image capturing in the JPEG format

Solder wires to the connection pads are mentioned as:

- Solder wire **1** (red) into the +5v connector pad
- Solder wire **2** (black) into the GND connector pad
- Solder wire **3** (white) into the TX connector pad
- Solder wire **4** (green) into the RX connector pad

Wiring the TTL Serial Camera for video capturing

The TTL Serial Camera board only supports NTSC video output and cannot be used as PAL. Now, solder two additional wires as shown in the following diagram:

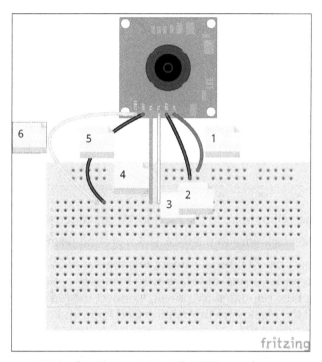

Wiring for video capturing with NTSC monochrome

Solder additional wires to the connection pads are mentioned as:

- Solder wire **5** (black) into the GND connector pad
- Solder wire **6** (yellow) into the CVBS connector pad

Testing NTSC video stream with video screen

Use an RCA jack and solder two wires, as stated here:

- Solder a yellow wire to the center terminal
- Solder a black wire to the outer terminal

Then make the other connections as follows:

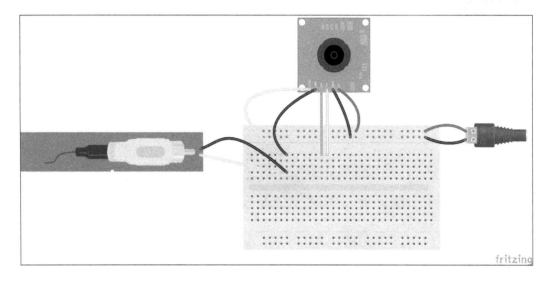

1. Connect the yellow wire to the RCA jack's signal terminal.

2. Connect the black wire to the RCA jack's ground terminal.

3. Now, connect the TTL serial camera to a regulated 5V power source, the red wire to positive, and the black wire to negative. You can use an Arduino board to get the regulated 5V power. Connect the red wire to Arduino 5V pin and the black wire to Arduino GND pin, and connect Arduino to the 9v power supply.

4. Finally, connect the soldered RCA jack to the NTSC monitor using an RCA video cable. If you are living in a region that does not support the NTSC broadcasting system, then you have to purchase a basic NTSC/PAL monitor. But some televisions support both NTSC and PAL broadcasting systems. Check your television's user manual for more information. If it does not support NTSC, then you have to purchase an NTSC- supported monitor from Adafruit (`http://www.adafruit.com/product/946`), or search eBay for a cheaper one.

Now, power up the monitor. You can see the monochrome video that has been captured by the TTL Serial Camera module. Next, we will move on to the most difficult part.

Connecting the TTL Serial Camera with Arduino and Ethernet Shield

Stack up your Arduino Ethernet Shield with the Arduino board as you did in the previous chapters and perform the following steps:

1. Connect your TTL Serial Camera module with the Arduino and Ethernet Shield as shown in the diagram below. Here, we will use two Arduino digital pins and a Software Serial port to communicate with the camera.

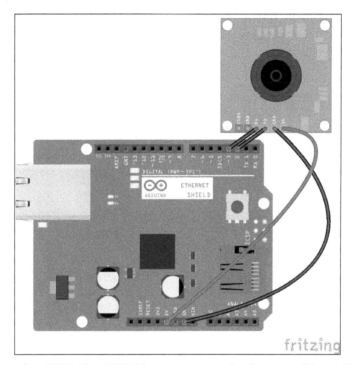

The Adafruit VC0706 Serial JPEG Camera is connected with Arduino Ethernet Shield

2. Connect camera TX to Arduino digital pin **2** and camera RX to Arduino digital pin **3**.

3. Connect camera **GND** to Arduino **GND** and camera **5V** to Arduino **5V**.

4. Now insert a Micro SD card into the SD card connector on the Ethernet shield. Remember the Arduino communicates with the SD card using digital pin **4**.

5. Download the Adafruit VC0706 camera library from GitHub by navigating to `https://github.com/adafruit/Adafruit-VC0706-Serial-Camera-Library`. After it has been downloaded, extract the ZIP file into your local drive.

6. Next, rename the folder `Adafruit-VC0706-Serial-Camera-Library` to `Adafruit_VC0706`, and move the renamed `Adafruit_VC0706` folder to the libraries folder. Note that the libraries folder resides in the Arduino IDE folder.

   Alternatively, in the recent version of Arduino IDE, you can add a new library ZIP file by navigating to **Sketch | Include Library | Add .ZIP Library...**. Then, browse the ZIP file and click on the **Open** button. This will add the particular library to the Arduino libraries folder.

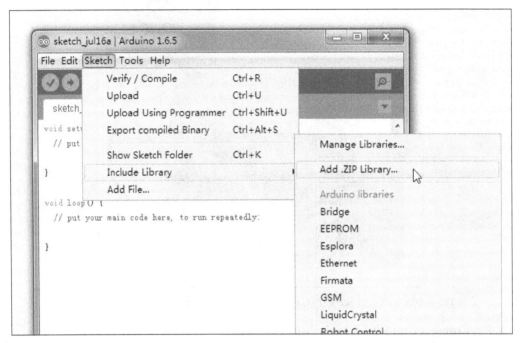

Including a new library by a ZIP file

7. Finally, restart the Arduino IDE.

The Adafruit VC0706 camera library includes sample sketches for image capturing and motion detection. You can verify them by navigating to **File | Examples | Adafruit VC0706 Serial Camera Library**.

Image capturing with Arduino

You can capture and save images in a Micro SD card using the Adafruit VC0706 camera library. Use the following steps to play with the sample sketches that ship with the Adafruit VC0706 camera library:

1. Open the Arduino IDE and go to **File** | **Example** | **Adafruit_VC0706** | **Snapshot**.

2. Upload the code on your Arduino board. (Also, you can copy the code B04844_04_01.ino from the Chapter 4 code folder)

3. Once uploaded, the camera will capture and save an image in the SD card with a resolution of 640 x 480. On the Arduino serial monitor, you can see some useful information about the image that is taken, such as the image resolution, image size in bytes, and the number of seconds taken to capture and save the image, which is in the JPG format.

Press the Arduino RESET button to capture the next image. You can use the RESET button to capture any subsequent images. But this sketch is limited to capturing images up to 100 times.

Insert your Micro SD card into the card reader of your PC, browse the SD card, and open the images using any image viewer installed in your computer. Cool!

Now, we will look into some important points of this sample code in the next section so that we can modify the code according to our requirements.

The Software Serial library

Arduino comes with hardware serial enabled, where pins 0 and 1 can be used to communicate with the serial devices. Pin 0 transmits (TX) the data to out and pin 1 receives (RX) data to in. However, using the Software Serial library, you can convert any digital pin in to TX or RX. For this project, we will use digital pins 2 for RX and 3 for TX.

```
SoftwareSerial(RX, TX)
```

The following code snippet shows how to use the Software Serial library to convert Arduino digital pin 2 as RX and digital pin 3 as TX:

```
// On Uno: camera TX connected to pin 2, camera RX to pin 3:
SoftwareSerial cameraconnection = SoftwareSerial(2, 3);
```

How the image capture works

The following code snippets show the important sections of the Arduino sketch.

To create a new object using the `Adafruit_VC0706` class, write a code line similar to the following. This will create the object `cam`:

```
Adafruit_VC0706 cam = Adafruit_VC0706(&cameraconnection);
```

The camera module can be used to take images in three different sizes. The largest size of the image we can take is 640 x 480. In the following code snippet, the camera will capture images in resolutions of 640 x 480. Uncomment the line you want to set as the image size.

```
// Set the picture size - you can choose one of 640x480, 320x240 or
160x120
// Remember that bigger pictures take longer to transmit!
cam.setImageSize(VC0706_640x480);  // biggest
//cam.setImageSize(VC0706_320x240);  // medium
//cam.setImageSize(VC0706_160x120);   // small
```

The `takePicture()` function of the `cam` object can be used to take a picture from the camera.

```
if (! cam.takePicture())
    Serial.println("Failed to snap!");
  else
    Serial.println("Picture taken!");
...
```

Then, create a filename for the new file by looking at the existing files stored in the SD card.

Finally, write the new image file on the SD card. This process is quite complicated and time consuming.

Uploading images to Flickr

Rather than saving the captured image in an SD card, we can automatically upload the image to Flickr. In the next section, we will learn how to do this with Flickr and Temboo cloud service.

Creating a Flickr account

Follow these steps to create a Flickr account:

1. Open your Internet browser, navigate to `https://www.flickr.com/`.

2. Click on the **Sign In** link in the top-right corner of the page. If you already have a Yahoo account, you can use the same login credentials to log in to Flickr.

3. If you don't have a Yahoo account, click on **Sign Up** in the top-left corner of the page, or on the **Sign up with Yahoo** button in the center of the page and follow the instructions to create a new Yahoo account.

4. After you have successfully logged in to Flickr, click on the **Explore** menu in the top, and in the resulting drop-down menu, click on **App Garden**.

5. **The App Garden** page will appear, as shown in the following screenshot:

Flickr: The App Garden page

6. Click on the **Create an App** link if it is not selected by default.

7. Under **Get your API Key**, click on **Request an API Key,** as shown in the following screenshot:

Flickr: The App Garden page

8. Click on the **APPLY FOR A NON-COMMERCIAL KEY** button, as shown here:

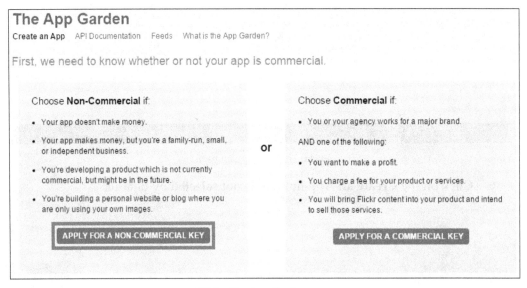

Flickr: The App Garden page

9. The **Tell us about your app** page will appear, as shown in the following screenshot:

Flickr: The App Garden page

10. Fill the following text boxes:
    ° **What's the name of your app?**: Give a short name for your app
    ° **What are you building?**: Give a brief description about your app and its purpose

11. Check the two checkboxes.

12. Click on the **SUBMIT** button.

13. The API **Key** and **Secret** for your new app will be displayed in the next page, as shown here:

Flickr: The App Garden page

Copy and paste the API Key and Secret into a notepad, if you think it will be easy for your reference later.

That's it for the moment. Later, you have to again visit the Flickr website, so don't sign out from Flickr. To access Flickr services, we have to create a Temboo account and make some configurations.

Creating a Temboo account

Temboo provides normalized access to 100+ APIs and databases. It provides code-based, task-specific code components called **Choreos** that can be used with the Arduino language to simplify the complex tasks such as uploading images to Flickr, sending SMS, sending Twitter tweets, and many more.

Let's look at how to create a new Temboo account, so you can use this account for experimenting with Temboo and Arduino.

1. First, navigate to `https://temboo.com/` using your Internet browser.

The Temboo home page

2. Then, you have to create a new user account in Temboo.

3. In the top-right corner of the page, there is a section for **Sign up**. Enter a name for your account, a valid e-mail address, and a password (which must have eight characters, at least one letter, and one number); agree with Temboo terms and click on the **Sign up** button. The **Welcome** page will appear, as shown here:

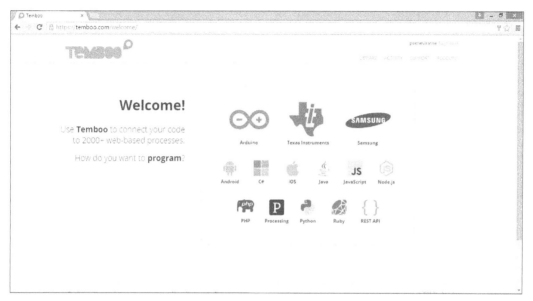

The Temboo Welcome page

Creating your first Choreo

Now, we are ready to create our first Choreo. To do this, we need to complete a series of configurations and processing steps with Temboo.

Initializing OAuth

In the top-right of the Temboo web page, click on **LIBRARY**. The **LIBRARY** page will appear. Under the **CHOREOS** pane (listed in the left-hand side of the page) go to **Flickr | OAuth** by clicking on the down arrow signs, and finally, click on **InitializeOAuth**.

First, enable **IoT Mode,** as shown in the following screenshot:

Enabling IoT mode

Then, configure the form as shown in the following steps:

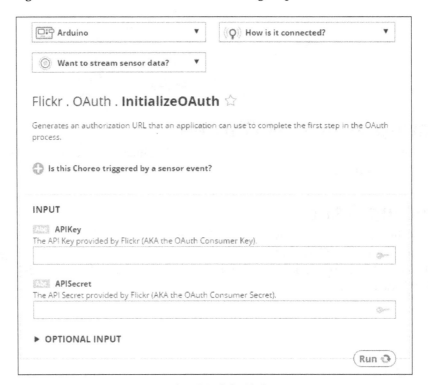

Initialize OAuth for Flicker

1. Select **Arduino** from the left drop-down menu. The default is **Arduino Yún**.

2. Select **Arduino Ethernet** from the **How is it connected?** drop-down menu. The **Tell us about your shield** dialog box will appear.

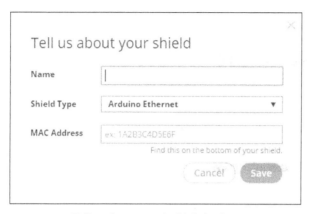

Tell us about your shield dialog box

3. Type a name for your shield and type the MAC address of your shield in the **MAC Address** field without any spaces. Then, click on the **Save** button.

4. Under the **INPUT** section, enter the following:

   ° **APIKey**: Enter the API key provided by Flickr

   ° **APISecret**: Enter the API Secret provided by Flickr

5. Click on the **Run** button to process the OAuth initialization. In a few seconds, the process will generate following output:

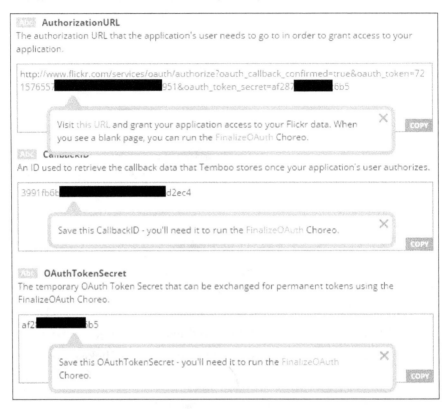

Output after the process of OAuth initialization for Flickr

The following listing of information is extracted from the preceding output. The information will differ according to your setup.

- ○ **Authorization URL**: `http://www.flickr.com/services/ oauth/authorize?oauth_callback_confirmed=true&oauth_ token=7215xxxxxxxxxxxxxxxxxxxxxxx951&oauth_token_ secret=af287xxxxxxxc6b5`

- ○ **CallbackID**: `3991fb6b-xxxxxxxxxxxxxxxx-b83e453d2ec4`

- ○ **OAuthTokenSecret**: `af287xxxxxxxc6b5`

6. Open a new browser tab and paste the authorization URL into the address bar and press the *Enter* key on your key board. A page will appear for authorization confirmation, as shown here:

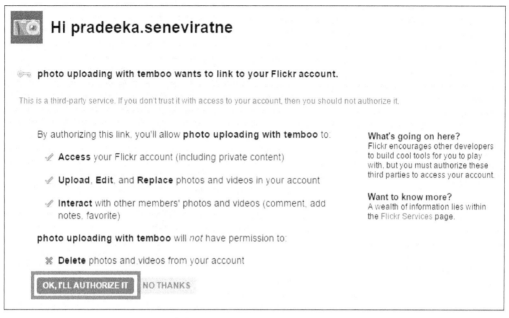

A Flicker user account authorization page

7. Click on the **OK, I'LL AUTHORIZE IT** button. Now, you will be navigated to a blank web page.

Finally, you have successfully authorized your app.

Finalizing OAuth

Perform the following steps to finalize OAuth:

1. Click on **FinalizeOAuth** after navigating to **Flickr | OAuth**. The **FinalizeOAuth** page will appear, as shown in the following screenshot:

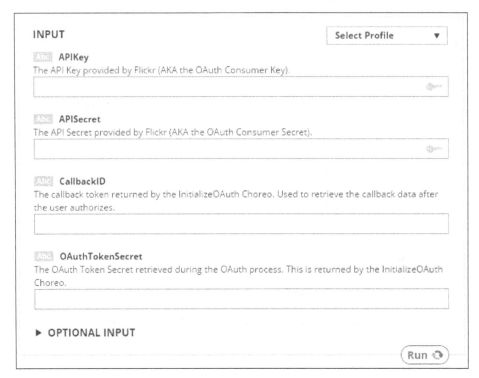

FinalizeOAuth for Flickr

2. Fill the following text boxes with the relevant information:
    -   **APIKey**: The API Key provided by Flicker for your app
    -   **APISecret**: The API secret provided by Flickr for your app
    -   **CallbackID**: The callback token returned by the InitializeOAuth process
    -   **OAuthTokenSecret**: The OAuth Token Secret retrieved during the OAuth process

3. Click on the **Run** button to process. Now you have finalized the OAuth process for your Flickr app.

Generating the photo upload sketch

In this section, you will learn how to generate the photo upload sketch. To achieve this, you need to perform the following steps:

1. Under **CHOREOS** go to **Flickr | Photos** and then click on **Upload**. The following screen will appear:

A Flicker photo upload Choreo

2. Fill the textboxes with the following information:
    ○   **AccessToken**: The Access Token retrieved during the OAuth process.
    ○   **AccessTokenSecret**: The AccessTokenSecret retrieved during the OAuth process.
    ○   **APIKey**: The API Key provided by Flickr.
    ○   **APISecret**: The API Secret provided by Flickr.
    ○   **ImageFileContents**: Keep this field blank.
    ○   **URL**: Any valid image URL. (for example, use `https://www.arduino.cc/en/uploads/Main/ArduinoEthernetFront450px.jpg`). Note that this specified image will be uploaded to your Flickr account for testing.

3. Click on the **Run** button to process the image upload to Flickr. If everything is correct, you will get a response, as shown in the following screenshot:

4. To verify the uploaded image, sign in to your Flickr account. On the Flickr web page, click go to **You | Camera Roll**. You can see the uploaded image by the Temboo cloud service, as shown here:

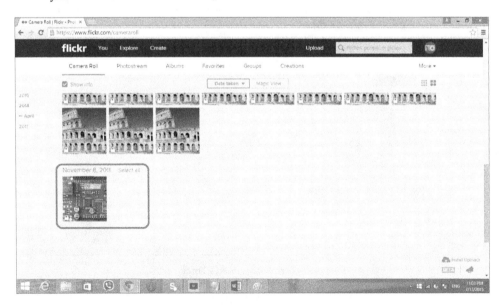

5. Continue with step 3, and scroll down the page. You can see two sections, **CODE** and **HEADER FILE**:

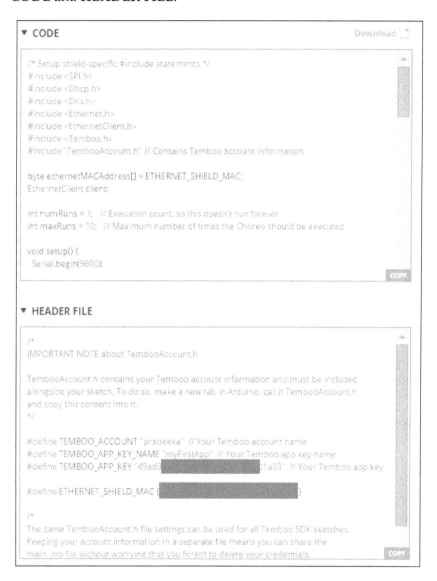

6. Now open a new Arduino IDE and copy and paste the generated code inside the **CODE** box.

7. Create a folder and rename it to **TembooAccount** inside your Arduino installation directory, and then under the **Libraries** folder. Copy the code inside the **HEADER FILE** box and paste it to a new Notepad file. Save the file as `TembooAccount.h` inside the `TembooAccount` folder.

8. Then, verify the code and upload it into your Arduino board.

9. Open the Arduino Serial Monitor. You can see the image upload status and it will upload your same image 10 times on Flickr. Open your Flickr's camera roll and verify the uploaded images.

10. The following Arduino sketch will upload an image (`https://www. arduino.cc/en/uploads/Main/ArduinoEthernetFront450px. jpg`) maximum of 10 times on Flickr. The sample sketch for this named `B04844_04_02.ino` can be copied from the `Chapter 4` code folder. Also, modify the API key values according to your Flickr and Temboo accounts.

The Arduino sketch for the `B04844_04_02.ino` file is:

```
/* Setup shield-specific #include statements */
#include <SPI.h>
#include <Dhcp.h>
#include <Dns.h>
#include <Ethernet.h>
#include <EthernetClient.h>
#include <Temboo.h>
#include "TembooAccount.h" // Contains Temboo account information

byte ethernetMACAddress[] = ETHERNET_SHIELD_MAC;
EthernetClient client;

int numRuns = 1;   // Execution count, so this doesn't run forever
int maxRuns = 10;  // Maximum number of times the Choreo should be
executed

void setup() {
  Serial.begin(9600);

  // For debugging, wait until the serial console is connected
  delay(4000);
  while(!Serial);

  Serial.print("DHCP:");
  if (Ethernet.begin(ethernetMACAddress) == 0) {
    Serial.println("FAIL");
    while(true);
  }
  Serial.println("OK");
  delay(5000);

  Serial.println("Setup complete.\n");
}
```

```
void loop() {
 if (numRuns <= maxRuns) {
  Serial.println("Running Upload - Run #" + String(numRuns++));

  TembooChoreo UploadChoreo(client);

  // Invoke the Temboo client
  UploadChoreo.begin();

  // Set Temboo account credentials
  UploadChoreo.setAccountName(TEMBOO_ACCOUNT);
  UploadChoreo.setAppKeyName(TEMBOO_APP_KEY_NAME);
  UploadChoreo.setAppKey(TEMBOO_APP_KEY);

  // Set Choreo inputs
  String APIKeyValue = "0c62beaaxxxxxxxxxxxxxxxxe3845ca";
  UploadChoreo.addInput("APIKey", APIKeyValue);
  String AccessTokenValue = "7215xxxxxxxxxxxxxxxxxxxxxx7b8e01";
  UploadChoreo.addInput("AccessToken", AccessTokenValue);
  String AccessTokenSecretValue = "d95exxxxxxxxfddb7";
  UploadChoreo.addInput("AccessTokenSecret",
AccessTokenSecretValue);
  String APISecretValue = "7277dxxxxxxxx7d696";
  UploadChoreo.addInput("APISecret", APISecretValue);
  String URLValue = "https://www.arduino.cc/en/uploads/Main/
ArduinoEthernetFront450px.jpg";
  UploadChoreo.addInput("URL", URLValue);

  // Identify the Choreo to run
  UploadChoreo.setChoreo("/Library/Flickr/Photos/Upload");

  // Run the Choreo; when results are available, print them to
serial
  UploadChoreo.run();

  while(UploadChoreo.available()) {
   char c = UploadChoreo.read();
   Serial.print(c);
  }
  UploadChoreo.close();
  }

  Serial.println("\nWaiting...\n");
  delay(30000); // wait 30 seconds between Upload calls
}
```

Connecting the camera output with Temboo

In the previous step, we successfully uploaded an image, which is in a remote server, on Flickr. Now, we are going to upload an image on Flickr, which is captured by the camera.

To do this, first we need to convert the image binary data stream to the base 64 stream.

Download the `base64.h` library from `https://github.com/adamvr/arduino-base64` and extract it inside to the `Libraries` folder.

Copy and paste the `B04844_04_03.ino` code from the sketches folder of this chapter and upload it on your Arduino board.

For every 30 seconds, your camera will capture an image and upload on Flickr.

Motion detection

Adafruit TTL serial camera has built-in motion detection capability. Using the VC0706 library, we can capture and upload the detected image to the Flickr. Here, we have used more similar code implementation as explained in the previous section of Motion Detection.

1. Open a new Arduino IDE and copy and paste the code `B04844_04_04.ino` from the `Chapter 4` code folder. Verify and upload the code on your Arduino board.

2. To test the motion, move an object in front of the camera. Wait nearly 30 seconds.

3. To verify the captured image, sign in to your Flickr account, and then, on the Flickr web page, go to **You | Camera Roll**. You can see the newly uploaded image by the Temboo cloud service.

Let's look at some important points in motion detection that are related to the Arduino sketch.

To enable the motion detection functionality on the VC0706 Camera module, you can use the following code line and set the parameter to `true`. The default is `false`. Note that the `cam` is the object of the `VC0706` class.

```
cam.setMotionDetect(true);
```

The motion is detected by the following function and it will return `true` when the motion is detected by the camera module.

```
cam.motionDetected();
```

Summary

Throughout this chapter, you learned how to build an Arduino security camera from scratch. Later, you can buy a dummy CCTV camera housing and secure your newly-built camera (Arduino and VC0706 Camera Module) by attaching it inside the housing. It will protect the electronic components from weather and any physical damages.

Further, you can modify the project with an Arduino WiFi shield or Cellular shield to make it wireless. Add a solar panel with a charger if you want to use it in a rural area that doesn't have electricity.

If you want more creativity, you can make a portable handheld camera for image capturing. Remember, you can use the Arduino **RESET** button to click!

In the next chapter, you will learn how to connect your Arduino to NearBus cloud using NearBus cloud connector, logging solar panel voltage data to the cloud, and displaying live data on the web browser.

5
Solar Panel Voltage Logging with NearBus Cloud Connector and Xively

Do you want to synchronize your Arduino board memory with cloud memory? Then this is the solution for memory mapping between Arduino and cloud. The memory mapping is done by mirroring or replicating a small part of Arduino's memory into the cloud's memory. So, reading or writing on the cloud's memory will have the same effect as reading or writing directly into the Arduino's memory.

The objective of this project is to log the voltage values generated by a solar cell against the time.

In this chapter, you will learn:

- About NearBus Cloud connector
- How to wire a solar cell with Arduino, and the use of the voltage divider
- How to install and use NearAgent with Arduino
- How to configure Xively with Arduino Ethernet
- How to combine NearBus with Xively
- How to display real time voltage logging with Xively
- How to write a simple HTML web page to display real time voltage logging that can be run on your mobile phone

Connecting a solar cell with the Arduino Ethernet board

We will use the following hardware to build the circuit:

- Arduino Ethernet board (https://www.sparkfun.com/products/11229) or Arduino UNO (https://www.sparkfun.com/products/11021), with Arduino Ethernet Shield (https://www.adafruit.com/products/201)
- A solar cell (https://www.sparkfun.com/products/7840)
- Two resistors (resistor values should be calculated on the open voltage of the solar cell); take a look at the *Building a voltage divider* section that follows for the calculation of values and color codes
- Some hook-up wires
- A 9V DC 650mA wall adapter power supply (https://www.sparkfun.com/products/10273)
- DC barrel jack adapter (https://www.sparkfun.com/products/10811)
- An Ethernet cable (https://www.sparkfun.com/products/8915)

Also, you will need a computer with an Arduino IDE installed.

Building a voltage divider

A voltage divider is a simple circuit that can be used to turn higher voltage into lower voltage through a series of two resistors. The resistor values depend on the input voltage and the mapped output voltage:

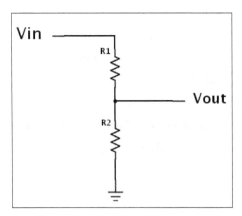

For this project, we are using Sparkfun Solar Cell Large — 2.5W (PRT07840). The open voltage of this solar cell is 9.15V (take a look at the datasheet for open voltage specification).

SparkFun Solar Cell Large - 2.5W Image courtesy of SparkFun Electronics (`https://www.sparkfun.com`)

So, we can calculate the resistor values for the voltage divider by using the following equation:

$$V_{out} = V_{in} \cdot \frac{R_2}{R_1 + R_2}$$

- V_{out} is 5V (the input voltage to Arduino)
- V_{in} is 9.15V (the output voltage from the solar cell)

Therefore, the following can be derived:

- $R1$ = 1200 Ohm = 1.2k (brown, red, red)
- $R2$ = 1500 Ohm = 1.5k (brown, green, red)

Building the circuit with Arduino

The following Fritzing diagram shows how to connect the voltage divider and solar cell with the Arduino Ethernet board. Now, start building the circuit according to the following diagram and steps provided:

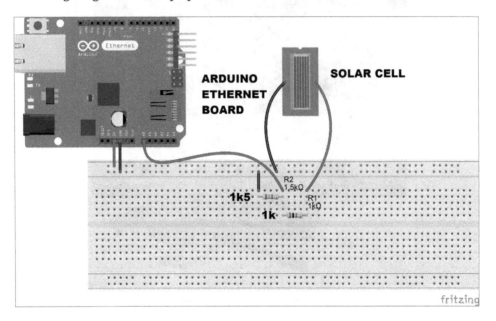

This particular solar cell comes with a DC barrel jack plug attached, and it is center positive. Plug it to the DC barrel jack adapter. Now solder two wires to the positive and negative terminals of the DC barrel jack adapter, as shown in the following image:

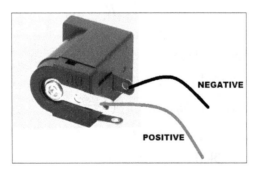

Connect the other wires as explained in the following steps:

1. Connect the voltage divider's output (V_{out}) with the Arduino analog pin 0 (A0).
2. Connect the solar cell's positive wire with voltage divider's V_{in}.
3. Connect the solar cell's negative wire to Arduino GND.
4. Connect the Arduino Ethernet board to a network switch or router using an Ethernet cable.
5. Power the Arduino Ethernet board using a 9V DC 650mA wall adapter power supply.

Now, the circuit and hardware setup is complete, and in the next section you will learn how to set up a NearBus account and connect your Arduino Ethernet shield to the NearBus Cloud for solar cell voltage logging.

Setting up a NearBus account

Setting up a NearBus account is simple. Visit the NearBus home page at `http://www.nearbus.net/` and click on **Sign Up** in the main menu. This will navigate you to the new user signup page with a simple form to enter your registration information. Enter your information as described in following steps:

1. **E-mail**: Type a valid e-mail address.
2. **User name**: Type your preferred name for the NearBus account.
3. **Password**: Type a secret word and don't share it with others.
4. Then, click on the checkbox of the captcha section to verify that you are a human.
5. Finally, click on the **Sign Up** button.

Now you have successfully registered with the NearBus website and you will be navigated to the **Login** page. Now, enter the following information to log in.

1. **Username**: Type your user name.
2. **Password**: Type your password.
3. Click on the **Login** button.

Defining a new device

Now, you can define a new device with the NearBus cloud connector. In this chapter, we will work with the Arduino Ethernet board. If you have an Arduino Ethernet Shield, you can stack it with an Arduino board and test it with the samples provided in this chapter.

1. On the NearBus website menu bar, click on **New Device**. You will be navigated to the **NEW DEVICE SETUP** page.

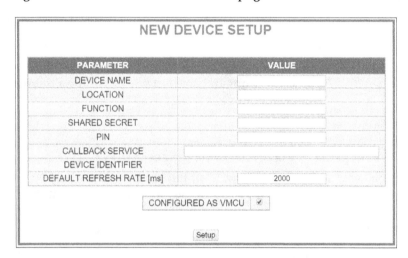

2. You can enter a value for each parameter and the only mandatory field is **SHARED SECRET**. It is eight characters long. Other fields are optional.

DEVICE NAME (Maximum 18 characters)	Arduino Ethernet
LOCATION	
FUNCTION	
SHARED SECRET	12345678
PIN	
CALLBACK SERVICE	
DEVICE IDENTIFIER	
DEFAULT REFRESH RATE [ms]	

3. Click on the **Setup** button.

Examining the device lists

After setting up the new device, you will navigate to the **DEVICE LIST** page. The NearBus system will assign a **DEVICE ID** to your new device and display your device name under the device alias. However, your new device will not have been mapped with NearBus. The mapped status shows as **DOWN,** which is highlighted in the following:

You will need this **DEVICE ID** when you write an Arduino sketch for this device.

Later, you can visit to the device list page by clicking on **Device List** on the menu bar.

Downloading the NearBus agent

To use your Arduino Ethernet Shield, or Arduino Ethernet board with the NearBus cloud connector, you must download and install the NearAgent code library. You can download the latest version of the NearBus library for Arduino at `http://www.nearbus.net/v1/downloads.html`. Also, you can visit the download page by clicking on **Downloads** on the NearBus web page menu bar. The following screenshot shows the **Download** page:

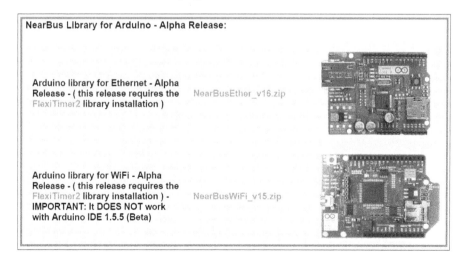

For this project, we need Arduino library for Ethernet, and the latest version is 16. Click on the **NearBusEther_v16.zip** link to download the library, or type `http://www.nearbus.net/downloads/NearBusEther_v16.zip` on your browser's address bar and hit *Enter* to download it on your computer's hard drive. Then, extract the downloaded ZIP file into the Arduino libraries folder.

Also, you need to download the FlexiTimer2 from `http://github.com/wimleers/flexitimer2/zipball/v1.1` and extract the ZIP file into the Arduino libraries folder. You can read more about the FlexiTimer2 at `https://github.com/wimleers/flexitimer2`, which is the GitHub page, and you can even download it from there.

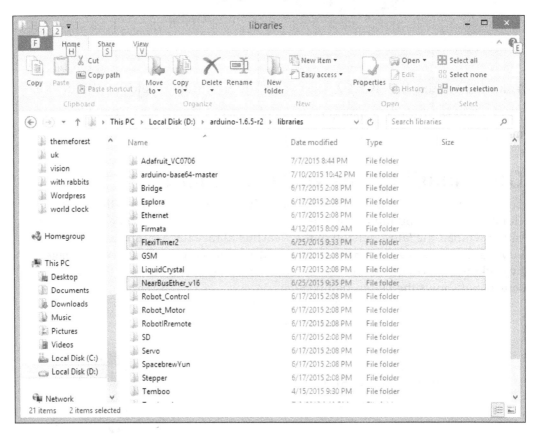

Perform the following steps to modify the sample code to read the voltage:

1. Open your Arduino IDE.

2. In the menu bar, click on **File | Examples | NearBusEther_v16 | Hello_World_Ether**. The sample code will load into the Arduino IDE. Also, you can copy and paste the sample sketch, `B04844_05_01.ino`, into your Arduino IDE which is located in the code folder of `Chapter 5`.

3. Save the sketch in another location by selecting **File | Save As** from the menu bar. Now, make the following modifications to the sample code to work with your Arduino Ethernet board or Ethernet Shield.

4. Modify the following code lines with your NearBus configuration's Device ID and Shared Secret. The Device ID can be found at the Device List page:

```
char deviceId []    = "NB101706"; // Put here the device_ID
generated by the NearHub ( NB1xxxxx )

char sharedSecret[] = "12345678"; // (IMPORTANT: mandatory
8 characters/numbers) - The same as you configured in the
NearHub
```

5. Replace the MAC address with your Arduino Ethernet board's MAC address:

```
byte mac[6] = { 0x90, 0xA2, 0xDA, 0x0D, 0xE2, 0xCD };  //
Put here the Arduino's Ethernet MAC
```

6. Comment the following line:

```
//pinMode(3, OUTPUT);
```

7. Then, uncomment the following line:

```
////////////////////////////////////
// Example 1 - Analog Input
// Mode: TRNSP
////////////////////////////////////
A_register[0] = analogRead(0); // PIN A0
```

Remember, our solar panel is connected to the Arduino analog pin, 0 (A0). But you can attach it to another analog pin and make sure that the pin number is modified in the sketch.

8. That's all. Now, connect your Arduino Ethernet board with the computer using an FTDI cable.

9. Select the board type as Arduino Ethernet (**Tools | Board | Arduino Ethernet**), and select the correct COM port (**Tools | Port**).

10. Verify and upload the sketch into your Arduino Ethernet board.

11. Now, revisit the **DEVICE LIST** page. You can see the Device's **STATE** is changed to **UP** and highlighted:

Now, your Arduino Ethernet Shield's internal memory is correctly mapped with the NearBus cloud.

In the next section, we shall learn how to feed our solar panel voltage readings to the Xively and display the real time data on a graph.

Creating and configuring a Xively account

Xively (formerly known as Cosm and Pachube) is a cloud-based platform that provides remote access to devices like Arduino and many more. You can read condensed information about Xively by visiting `https://en.wikipedia.org/wiki/Xively`.

Using your web browser, visit `http://xively.com/`:

There is no link label to sign up in the web page, so type `https://personal.xively.com/signup` in your web browser's address bar and directly visit the **Sign Up** page.

After a successful sign up with Xively, you will get an e-mail with an activation link. Click on the activation link.

You will be redirected to the **Welcome to Xively** web page:

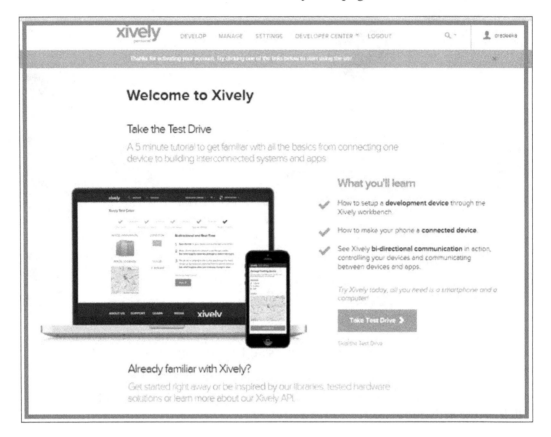

Click on **DEVELOP** from the top menu bar. The **Development Devices** page will appear:

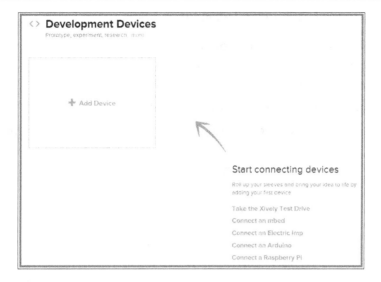

Click on **+Add Device**. The **Add Device** page will appear:

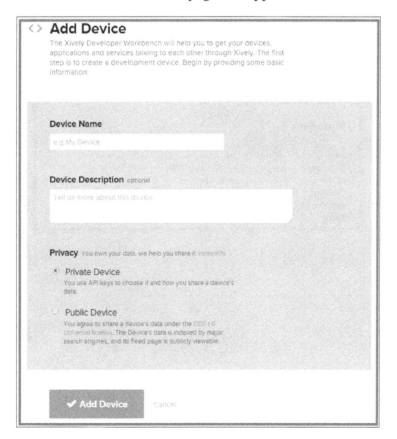

Fill the following textboxes with relevant information:

- **Device Name**: Give a name for your device, for example, Voltage Logger.
- **Device Description**: Give a brief description of your device, for example, Logging solar panel voltage.
- Click on the **Private Device** option radio button.
- Click on the **Add Device** button. The following web page will appear:

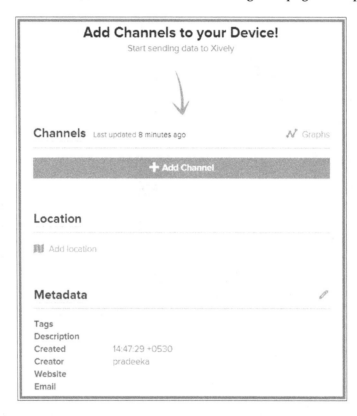

Click on the **+Add Channel** button:

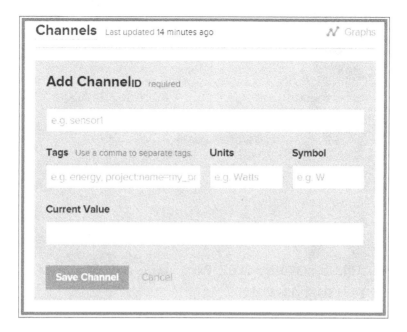

To add a channel, follow these steps:

1. Fill in the following information:
    - ° **Channel ID**: 1 (but you can use any name, for example, `sensor1` or `logger1`)
    - ° **Units**: Volts
    - ° **Symbol**: V
    - ° **Current Value**: Leave it blank or type `0`

2. Click on the **Save Channel** button to save the channel.

3. On the right-hand side of the page, under **API Keys,** you can find out the Xively **API Key** and **Feed ID** for this device.

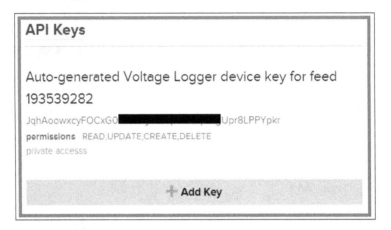

- ○ **API Key: GE0sSoyHziZ3Pxxxxxxxxxxxxxxqb7adMUA5yaVUu5psjs**
- ○ **Feed ID: 1913539282**

Configuring the NearBus connected device for Xively

Perform the following steps to configure the NearBus connected device for Xively:

1. Log in to your NearBus account.
2. Click on the **DEVICE LIST** menu. The **DEVICE LIST** page will appear:

3. From the drop-down menu, select **COSM CONFIG**.

4. Click on the **Setup** button. The **COSM CONNECTOR (xively.com)** page will appear:

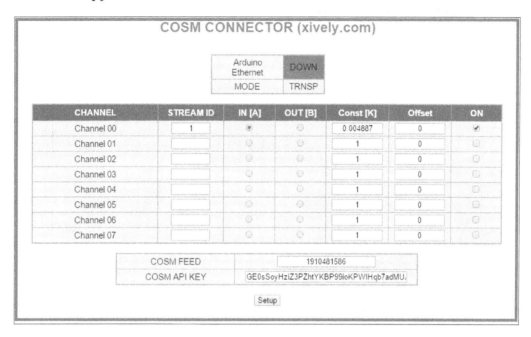

5. In this page, you have to enter some configuration settings in order to work your Arduino with NearBus and Xively. Here, we are using the NearBus channel 0 to communicate with our device. So, configure the following entries for the channel 0 entry using the **CMOS CONNECTOR (xively.com)** page:

   ° **STREAM ID**: 1.

   ° **IN[A]**: Click to enable.

   ° **Const[K]**: 0.004887 (Arduino analog input can accept values from 0-1023, so we need to map the input voltage (in this case 5v) with it). Divide 5V by 1024, then you will get 0.0048828125. Copy and paste it to the **Const[K]** textbox.

   ° **Offset**: 0.

   ° **ON**: Click on the checkbox to check.

   ° **COSM FEED**: Type the Feed ID generated by the Xively.

   ° **COSM API KEY**: Type the API Key generated by the Xively.

6. Click on the **Setup** button.

7. Now, switch to your Xively web page. On the left-hand side of the page, you can see a graph displaying your solar cell voltage with time. If you can't see the graph, click on the channel ID to expand the graph:

Developing a web page to display the real-time voltage values

Use the following chart to find the time zone for your device. The following two examples will show you how to find the correct time zone value:

- **Example 1:** Let's assume your device is located in Sri Jayawardenepura, the time zone is 5.5, that is, UTC +05:30 (Chennai, Kolkata, Mumbai, New Delhi, Sri Jayawardenepura)

- **Example 2**: If it is located in Newfoundland, the time zone is -3.5, that is, UTC -03:30 Newfoundland

Zone	Place names
UTC -11:00	International Date Line West, Midway Island, Samoa
UTC -10:00	Hawaii
UTC -09:00	Alaska
UTC -08:00	Pacific Time (US & Canada), Tijuana
UTC -07:00	Arizona, Chihuahua, Mazatlan, Mountain Time (US & Canada)
UTC -06:00	Central America, Central Time (US & Canada), Guadalajara, Mexico City, Monterrey, Saskatchewan
UTC -05:00	Bogota, Eastern Time (US & Canada, Indiana (East)), Lima, Quito
UTC -04:30	Caracas
UTC -04:00	Atlantic Time (Canada), La Paz, Santiago
UTC -03:30	Newfoundland
UTC -03:00	Brasilia, Buenos Aires, Georgetown, Greenland
UTC -02:00	Mid-Atlantic
UTC -01:00	Azores, Cape Verde Is.
UTC +00:00	Casablanca, Dublin, Edinburgh, Lisbon, London, Monrovia
UTC +01:00	Amsterdam, Belgrade, Berlin, Bern, Bratislava, Brussels, Budapest, Copenhagen, Ljubljana, Madrid, Paris, Prague, Rome, Sarajevo, Skopje, Stockholm, Vienna, Warsaw, West Central Africa, Zagreb
UTC +02:00	Athens, Bucharest, Cairo, Harare, Helsinki, Istanbul, Jerusalem, Kyev, Minsk, Pretoria, Riga, Sofia, Tallinn, Vilnius
UTC +03:00	Baghdad, Kuwait, Moscow, Nairobi, Riyadh, St. Petersburg, Volgograd
UTC +03:30	Tehran
UTC +04:00	Abu Dhabi, Baku, Muscat, Tbilisi, Yerevan
UTC +04:30	Kabul
UTC +05:00	Ekaterinburg, Islamabad, Karachi, Tashkent
UTC +05:30	Chennai, Kolkata, Mumbai, New Delhi, Sri Jayawardenepura
UTC +05:45	Kathmandu
UTC +06:00	Almaty, Astana, Dhaka, Novosibirsk
UTC +06:30	Rangoon
UTC +07:00	Bangkok, Hanoi, Jakarta, Krasnoyarsk
UTC +08:00	Beijing, Chongqing, Hong Kong, Irkutsk, Kuala Lumpur, Perth, Singapore, Taipei, Ulaan Bataar, Urumqi

Zone	Place names
UTC +09:00	Osaka, Sapporo, Seoul, Tokyo, Yakutsk
UTC +09:30	Adelaide, Darwin
UTC +10:00	Brisbane, Canberra, Guam, Hobart, Melbourne, Port Moresby, Sydney, Vladivostok
UTC +11:00	Magadan, New Caledonia, Solomon Is.
UTC +12:00	Auckland, Fiji, Kamchatka, Marshall Is., Wellington
UTC +13:00	Nuku'alofa

Displaying data on a web page

Now, we will look at how to display the temperature data on a web page using HTML and JavaScript by connecting to the Xively cloud.

1. Copy the following `index.html` file from the code folder of `Chapter 5` to your computer's hard drive.

2. Using a text editor (Notepad or Notepad++), open the file and edit the highlighted code snippets using your NearBus and Xively device configuration values. Modify the device ID, user, and password:

```
var device_id = "NB101706"; // Your device ID
var user = "****"; // Your NearBus Web user
var pass = "****"; // Your NearBus Web password
```

Following are the variables:

   ◦ `device_id`: Your NearBus device ID

   ◦ `user`: Your NearBus user name

   ◦ `pass`: Your NearBus password

3. Replace `1` with your NearBus channel ID:

```
ret = NearAPIjs( "ADC_INPUT", device_id , 1, 0, "RONLY" );
```

4. Replace `1910481586` with your Xively device Feed ID, `1.png` with your NearBus channel ID (only replace the number part) and `5.5` with your time zone.

```
<div id="div_temp_chart_cm"> <img
src="https://api.cosm.com/v2/feeds/1910481586/datastreams/1
.png?width=750&height=400&colour=%23f15a24&duration=3hours&
show_axis_labels=true&detailed_grid=true&timezone=5.5" >
</div>
```

You can also modify the cosm to Xively in the preceding URL because both are working. The modified URL can be written as follows:

```
<div id="div_temp_chart_cm"> <img
src="https://api.xively.com/v2/feeds/1910481586/datastreams
/1.png?width=750&height=400&colour=%23f15a24&duration=3hour
s&show_axis_labels=true&detailed_grid=true&timezone=5.5" >
</div>
```

5. Now, save and close the file. Then, open the file using your preferred web browser. You will see a graph displaying the real-time voltage values against the time which is continuously updating.

6. Also, you can copy the file into your smart phone's SD card or its internal memory, and then open it with the mobile web browser to see the real time graph.

7. The following image shows a real time graph that is plotting the output voltage of a solar cell, where the x axis represents the time (**t**) and the y axis represents the voltage (**V**):

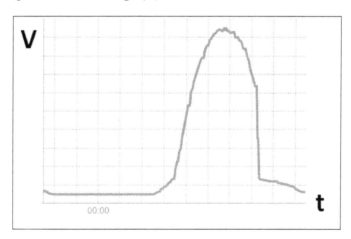

Summary

In this chapter, you have learned how to log your solar cell voltage using the NearBus and Xively cloud platforms and access them remotely from anywhere in the world using a mobile device. You can modify this project to log data from any type of sensor and also add more channels to display multiple data streams on a single graph.

In the next chapter, you will learn how to work with GPS and combine it with the Internet. Also, you will learn how to plot locations using Google Maps.

6

GPS Location Tracker with Temboo, Twilio, and Google Maps

Location tracking is important when you want to find the exact location of movable objects, such as vehicles, pets, or even people. GPS technology is very helpful in getting precise locations, which makes it possible to create real-time tracking devices.

In this chapter you will learn:

- How to connect the Arduino GPS shield with the Arduino Ethernet board
- How to install and use TinyGPSPlus library with the Arduino Ethernet board
- How to extract location data and time with Arduino GPS shield in conjunction with TinyGPSPlus library
- About Google Maps JavaScript API that displays the current location on Google Maps with GPS data
- How to get GPS location data by SMS with Twilio and Temboo

Hardware and software requirements

You will need the following hardware and software to complete this project:

Hardware requirements

- Arduino Ethernet board (`https://www.sparkfun.com/products/11229`)
- SparkFun GPS Shield kit (`https://www.sparkfun.com/products/13199`)
- FTDI Cable 5V (`https://www.sparkfun.com/products/9718`)
- 9V DC 650mA wall adapter power supply (`https://www.sparkfun.com/products/10273`)

Software requirements

- TinyGPSPlus library (`https://github.com/mikalhart/TinyGPSPlus/archive/master.zip`)

Getting started with the Arduino GPS shield

Arduino GPS shield lets your Arduino board receive information from the **GPS (Global Positioning System)**. The GPS is a satellite-based navigation system made up of a network of 24 satellites.

Arduino GPS shield consists of a GPS receiver that can be used to receive accurate time signals from the GPS satellite network and calculate its own position.

Arduino GPS Shield is currently manufactured by various electronics kit suppliers. The most popular manufacturers are SparkFun Electronics and Adafruit. Throughout this project, we will use the SparkFun GPS Shield kit (`https://www.sparkfun.com/products/13199`).

The kit comes with an EM-506 GPS module and the Arduino stackable header kit. Click on **Assembly Guide** (`https://learn.sparkfun.com/tutorials/gps-shield-hookup-guide`) in the product page, and follow the instructions to solder the headers and GPS module to the shield.

The Arduino GPS shield kit: Image taken from SparkFun Electronics Image courtesy of SparkFun Electronics (https://www.sparkfun.com)

Connecting the Arduino GPS shield with the Arduino Ethernet board

To connect the Arduino GPS shield with the Arduino Ethernet board, perform the following steps:

1. Stack your Arduino GPS shield with the Arduino Ethernet board.

2. Move the **UART/DLINE** switch to the **DLINE** position. This is a two-way switch that can be used to select the **UART** or **DLINE** mode to communicate GPS shield with Arduino.

   ° **UART**: This connects the GPS module's serial lines to Arduino hardware serial (D0/RX and D1/TX).

   ° **DLINE**: This connects the GPS module's serial lines to the Arduino software serial (D2 and D3). See the solder marks label next to the UART/DLINE switch.

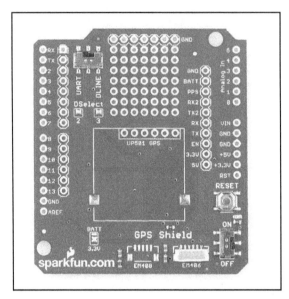

The Arduino GPS shield PCB: Image courtesy of SparkFun Electronics (https://www.sparkfun.com)

3. Connect the 9V DC power supply to your Arduino Ethernet board. Then, connect the Arduino Ethernet board to the computer with an FTDI cable a USB to Serial (TTL Level) converter.

4. Now, download the TinyGPSPlus library from https://github.com/mikalhart/TinyGPSPlus/archive/master.zip and extract it to your Arduino Installation's libraries folder.

Testing the GPS shield

Follow these steps to test the GPS shield:

1. Open a new Arduino IDE, then copy and paste the sample code B04844_06_01.ino from the Chapter 6 code folder of this book. (Note that this is the sample code included with the TinyGPSPlus library to display the current location by latitude and longitude with date and time). You can also open this sketch by navigating to **File | Examples | TinyGPSPlus | DeviceExample** on the menu bar.

2. Verify and upload the sketch to your Arduino board or Arduino Ethernet board.

3. Open the Arduino Serial Monitor by going to **Tools | Serial Monitor**. The following output will be displayed:

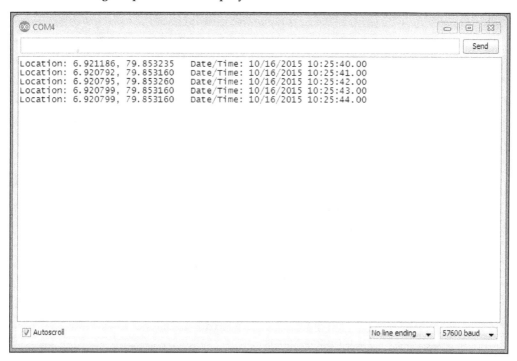

In each entry, the current location is displayed with latitude, longitude, and date/time. Next, you will learn how to use these values to display the location on Google Maps.

Displaying the current location on Google Maps

Google Maps JavaScript API can be used to display the current location with a marker on Google Maps. We can simply pass the latitude and longitude to the Google JavaScript API library and display the current location as a simple marker.

The following steps will explain you how to display the Arduino GPS shield's current location on Google Maps:

1. Open a new Arduino IDE and paste the sample code `B04844_06_02.ino` from the `Chapter 6` code folder. Verify and upload the sketch to your Arduino Ethernet Shield or Arduino Ethernet board.

2. The code consists a mix of Arduino, HTML, and JavaScript. Let's look at some important points of the code.

   ○ The following JavaScript function creates a new Google map position with latitude and longitude:

   ```
   var myLatlng = new google.maps.LatLng(-
   25.363882,131.044922);
   ```

   The latitude and longitude values should be replaced with the real time returning values of the Arduino GPS shield as follows:

   ```
   var myLatlng = new
   google.maps.LatLng(gps.location.lat(),gps.location.lng());
   ```

   ○ The following JavaScript function will create a map and display a simple marker on Google Maps based on the location provided by the map options:

   ```
   var map = new google.maps.Map(document.getElementById('map-
   canvas'), mapOptions);
   ```

3. Open your web browser and type the IP address of the Arduino Ethernet Shield and navigate to the site. Read *Chapter 1, Internet-Controlled PowerSwitch,* for information on how to find the IP address of your Arduino Ethernet Shield. Example: `http://192.168.10.177`.

4. Your web browser will display your GPS shield's current location on the Google Map, as shown in the following screenshot:

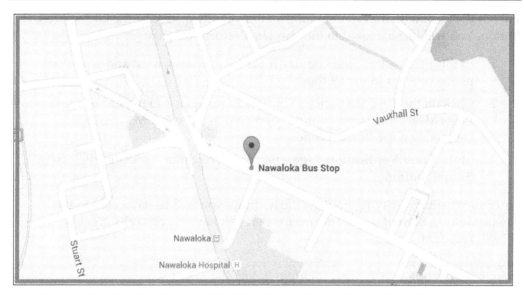

The current location of the Arduino GPS shield is displayed on the Google Map with a marker icon

In the next section, you will learn how to send the current GPS location by SMS to the client using Twilio and Temboo.

Getting started with Twilio

The Twilio platform provides API to programmatically send, receive, and track SMS messages worldwide, while also letting you test your SMS-enabled applications within a sandbox environment.

Creating a Twilio account

Using your Internet browser, visit `https://www.twilio.com/`. The Twilio landing page will be displayed with the sign up and log in options.

Click on the **SIGN UP** button. You will be navigated to the new user registration form.

Fill out the form with the relevant information and then click on the **Get Started** button. You will be navigated to the **Human Verification** page:

1. In this page, select your country from the drop-down list and type your phone number in the textbox.

2. Click on the **Text Me** button. You will be navigated to the **Enter Verification Page**. Meanwhile, your mobile phone will receive an SMS message containing a verification code.

3. In the **Enter Verification Page**, enter the verification code and click on the **Submit** button.

After successfully verifying your Twilio account, you will be navigated to the Twilio **Getting Started** page. This means that you have successfully created a Twilio account.

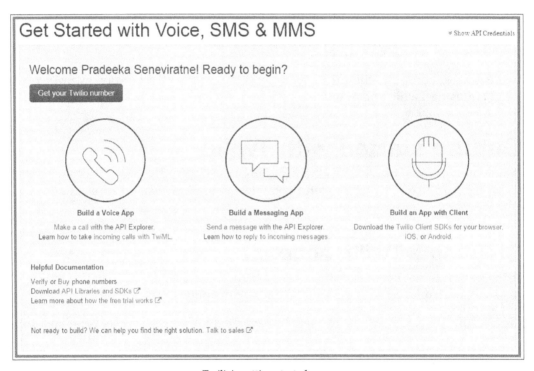

Twilio's getting started page

Finding Twilio LIVE API credentials

At the top of the Twilio getting started page, you will find the **Show API Credentials** link. Click on it, and the **API Credentials** panel will expand and display the following information:

- **ACCOUNT SID**
- **AUTH TOKEN**

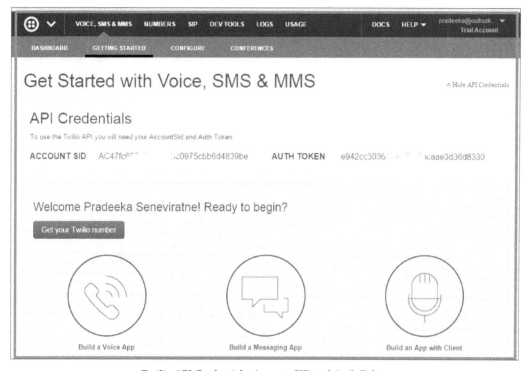

Twilio API Credentials: Account SID and Auth Token

You will need the **ACCOUNT SID** and **AUTH TOKEN** in the next section when you connect your Twilio account with Temboo. However, the default account type is a trial account with limited API calls allocated. If you want to get the full benefit of Twilio, upgrade the account.

Finding Twilio test API credentials

At the top-right of the page, click on your account name, and from the drop-down menu, click on **Account**. The **Account Settings** page will appear, as shown here:

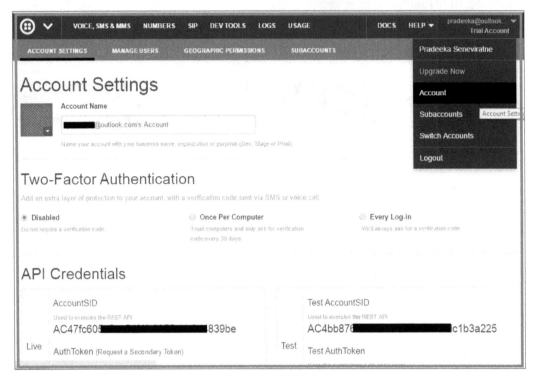

Twilio test API credentials: Test Account SID and Test Auth Token

In this page, under **API Credentials,** you can find the **Test Account SID** and **Test Auth Token** to test your apps with Twilio.

Get your Twilio number

Your Twilio account provides phone numbers to use with Voice, SMS, and MMS. You can obtained one such number by following these instructions:

1. Click on the **Voice, SMS & MMS** menu item at the top of the page.

2. Click on **GETTING STARTED** in the submenu of **Voice, SMS & MMS**.

3. Click on the **Get your Twilio Number** button. Your first Twilio phone number will generate and you can choose the number by clicking on the **Choose this number** button. Also, you can search for a different number by clicking on Search, for a different number link.

   Some countries, such as Australia, do not have SMS capability for trial accounts. Use a United States number, which will enable you to send SMS internationally.

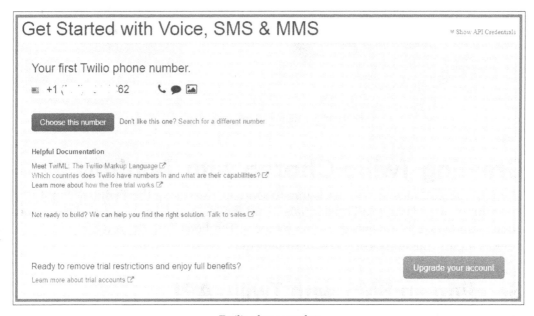

Twilio phone number

4. The following page will be displayed as a confirmation. You can further configure your Twilio phone number by clicking on the **Configure your number** button:

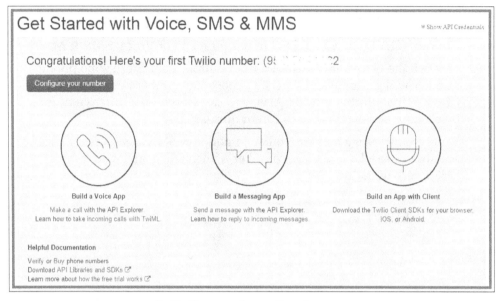

Twilio Phone number configuration page

Creating Twilio Choreo with Temboo

The Temboo provides us with a Choreo to send an SMS using the Twilio account. This Choreo uses Twilio API credentials to authenticate and send SMS to destination phone numbers. The advantage is that by using Temboo Choreos, you can write more complex functions using few lines of code.

Sending an SMS with Twilio API

To send an SMS with Twilio API, perform the following steps:

1. Sign in to your Temboo account. If you still don't have a Temboo account, create one as discussed in *Chapter 5, Solar Panel Voltage Logging with NearBus Cloud Connector and Xively.*

2. Under **CHOREOS**, expand **Twilio,** then expand **SMSMessages** and click on **SendSMS**.

3. The right-hand side of the page will load the Twilio SendSMS configuration form.

4. Turn **ON** the **IoT Mode**.

Twilio SendSMS form

5. Fill out the following textboxes with your Twilio API settings:
    ○ **AccountSID**: Type the Twilio Test Account SID
    ○ **AuthToken**: Type the Twilio Test AuthToken
    ○ **Body**: You can type any text message here in order to test
    ○ **From**: Type Twilio Sandbox number (use your Twilio phone number)
    ○ **To**: The destination phone number (use your phone number that associated with the Twilio account)

6. Click on the **Run** button to send the SMS to your phone.

Send a GPS location data using Temboo

To send a GPS location data using Temboo, perform the following steps:

1. Open a new Arduino IDE and copy and paste the sketch `B04844_06_03.ino` from the `Chapter 6` code folder.

2. Replace the `ToValue` and `FromValue` phone numbers, as shown here:

```
String ToValue = "+16175XXX213";
SendSMSChoreo.addInput("To", ToValue);
String FromValue = "+16175XXX212";
SendSMSChoreo.addInput("From", FromValue);
```

3. Save the `B04844_06_03.ino` sketch in your local drive inside a new folder. Copy the code generated in the `HEADER FILE` section under the `Twilio SendSMS` section, and paste it into a new Notepad file. Save the file as `TembooAccount.h` in the same location.

4. Verify the sketch. If you get a compiler error indicating that the `TembooAccount.h` header file is missing, restart the Arduino IDE and open the `B04844_06_03.ino` sketch again and then verify. This will probably solve the issue.

5. Upload the sketch into your Arduino Ethernet board.

6. You will receive the first SMS including the GPS location data from your device. Wait for 30 minutes. You will receive the second SMS. You can change the delay between SMS messages by modifying the following code line as shown:

```
delay(1800*1000); // wait 30 minutes between SendSMS calls
```

The value 1,800 seconds is equal to 30 minutes. To convert the 1,800 seconds into milliseconds, multiply it by 1,000.

Summary

In this chapter, you learned how to connect the Arduino GPS shield with Arduino Ethernet Shield while displaying the current location using Google maps with Google Maps JavaScript API. You also used Twilio and Temboo APIs to send SMS messages with GPS location data to the user.

In the next chapter, you will learn how to build a garage door light that can be controlled using Twitter tweets with the combination of Python and Python-Twitter (a Python wrapper around the Twitter API).

7
Tweet-a-Light – Twitter-Enabled Electric Light

In *Chapter 1, Internet-Controlled PowerSwitch*, we learned how to control a PowerSwitch Tail (or any relay) through the Internet by using the Arduino Ethernet library. Now, we will look into how Twitter tweets can be used to control the PowerSwitch Tail.

In this chapter, we will learn:

- How to install Python on Windows
- How to install some useful libraries on Python, including pySerial and Tweepy
- How to create a Twitter account and obtain Twitter API keys
- How to write a simple Python Script to read Twitter tweets and write data on serial port
- How to write a simple Arduino sketch to read incoming data from serial port

Hardware and software requirements

To complete this project, you will require the following hardware and software.

Hardware

- Arduino UNO Rev3 board (`https://store.arduino.cc/product/A000066`)
- A computer with Windows installed and Internet connected

- PowerSwitch Tail (120V or 240V depending on your voltage of mains electricity supply) – (`http://www.powerswitchtail.com/Pages/default.aspx`)
- A light bulb (120V or 240V depending on your voltage of mains electricity supply), holder, and wires rating with 120V/240V
- A USB A-to-B cable (`https://www.sparkfun.com/products/512`)
- Some hook-up wires

Software

The software needed for this project is mentioned under each topic so that it will be easier to download and organize without messing things up.

Getting started with Python

Python is an interpreted, object-oriented, and high-level computer programming language with very powerful features that's easy to learn because of its simple syntax. For this project, we can easily write an interface between Twitter and Arduino using the Python script.

Installing Python on Windows

The following steps will explain how to install Python on a Windows computer:

1. Visit `https://www.python.org/`.
2. Click on **Downloads | Windows**.

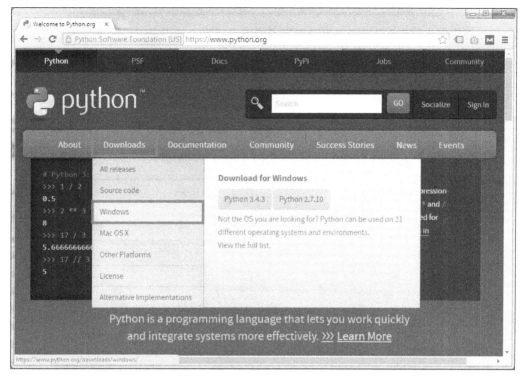

The Python home page

3. Then, you will navigate to the **Python Releases for Windows** web page:

The Python download page

4. Python can be downloaded from two development branches: legacy and present. The legacy releases are labeled as 2.x.x, and present releases are labeled as 3.x.x. (For reference, the major difference of 2.7.x and 3.0 can be found at `http://learntocodewith.me/programming/python/python-2-vs-python-3/`). Click on the latest (see the date) Windows x86-64-executable installer to download the executable installer setup file to your local drive under Python 3.x.x.

5. Alternately, you can download a web-based installer or embedded ZIP file to install Python on your computer.

6. Browse the default `Downloads` folder in your computer and find the downloaded setup file. (My default downloads folder is `C:\Downloads`).

7. Double-click on the executable file to start the setup:

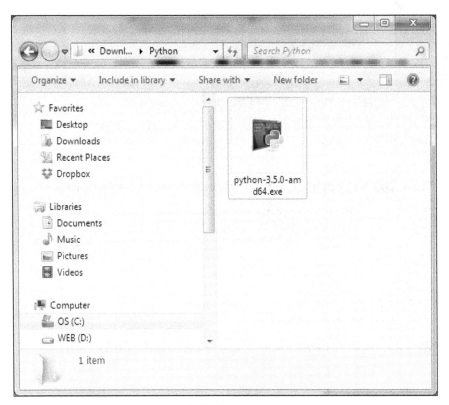

The Python setup

8. Click on the **Run** button if prompted as a security warning:

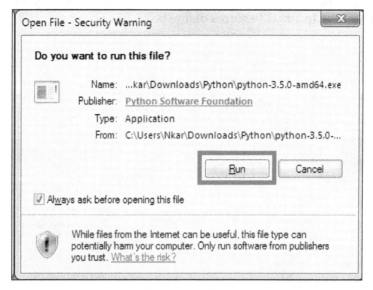

Security warning

9. The Python setup wizard starts:

The Python setup wizard – start screen

10. Optionally, you can check **Add Python 3.5 to PATH**, or later, you can add it using Windows system properties. Click on the **Customize installation** section. The **Optional Features** dialog box will appear:

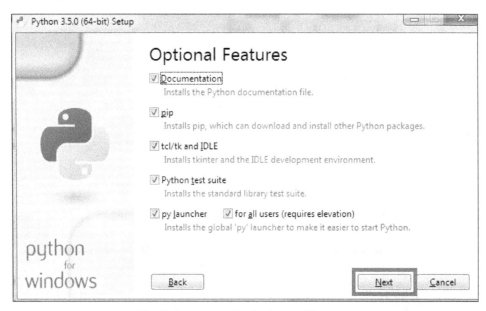

The Python setup wizard — Optional Features

11. Click on the **Next** button to proceed. The **Advanced Options** dialog box will appear. Keep the selected options as default.

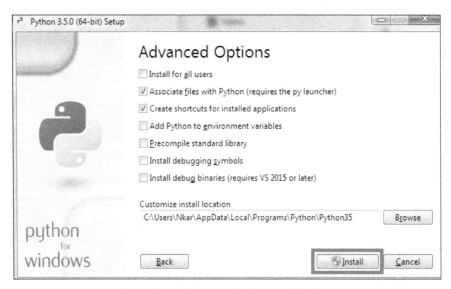

The Python setup wizard — Advanced Options

12. The default installation path can be found under **Customize install location**. If you like, you can change the installation location by clicking on the **Browse** button and selecting a new location in your computer's local drive.

13. Finally, click on the **Install** button.

14. If prompted for User Access Control, click on **OK**. The **Setup Progress** screen will be displayed on the wizard:

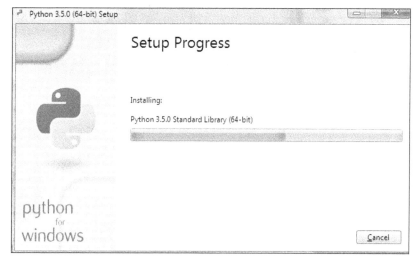

Python setup installation progress

15. If the setup is successful, you will see the following dialog box. Click on the **Close** button to close the dialog box:

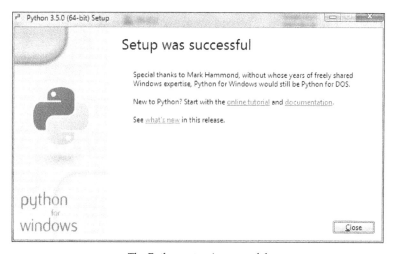

The Python setup is successful

Setting environment variables for Python

If you have already set to **Add Python 3.5 to PATH** for writing the environment variables during the Python setup installation process, ignore this section. If not, then follow these steps to set environment variables for Python.

1. Open the Windows **Control Panel** and click on **System**. Then, click on **Advanced system settings**.

2. The **System Properties** dialog box will appear. Click on the **Advanced** tab. Then, click on the **Environment Variables...** button:

The System Properties dialog box

3. The **Environment Variables** dialog box will appear. Click on the **New…** button under user variables:

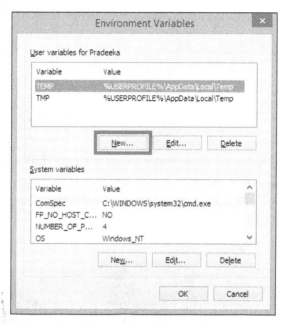

The Environment Variables dialog box

4. The **New User Variable** dialog box appears:

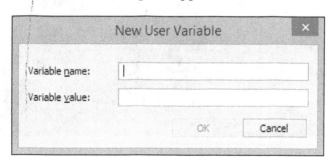

The New User Variable dialog box

5. Type the following for the respective textboxes:

    ○   **Variable name:** PATH

    ○   **Variable Value:** C:\Users\Pradeeka\AppData\Local\Programs\
        Python\Python35;C:\Users\Pradeeka\AppData\Local\
        Programs\Python\Python35\Lib\site-packages\;C:\Users\
        Pradeeka\AppData\Local\Programs\Python\Python35\
        Scripts\;

Modify the preceding paths according to your Python installation location:

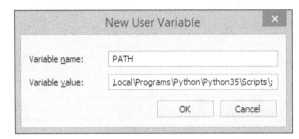

The New User Variable dialog box

6. Click on the **OK** button three times to close all the dialog boxes.

7. Open Windows Command Prompt and type `python`, and then press the *Enter* key. The Python Command Prompt will start. The prompt begins with >>> (three greater than marks):

```
Command Prompt - python                                  _ □ ×

Microsoft Windows [Version 6.3.9600]
(c) 2013 Microsoft Corporation. All rights reserved.

C:\Users\Pradeeka>python
Python 3.5.0b4 (v3.5.0b4:c0d641054635, Jul 26 2015, 07:11:12) [MSC v.1900 64 bit
 (AMD64)] on win32
Type "help", "copyright", "credits" or "license" for more information.
>>> _
```

Python Command Prompt

This ensures that the Python environment variables are successfully added to Windows. From now, you can execute Python commands from the Windows command prompt. Press *Ctrl + C* to return the default (Windows) command prompt.

Installing the setuptools utility on Python

The setuptools utility lets you download, build, install, upgrade, and uninstall Python packages easily. To add the setuptools utility to your Python environment, follow the next steps. At the time of writing this book, the setuptools utility was in version 18.0.1.

1. Visit the setuptools download page at https://pypi.python.org/pypi/ setuptools.

2. Download the ez_setup.py script by clicking on the link (https://bootstrap.pypa.io/ez_setup.py):

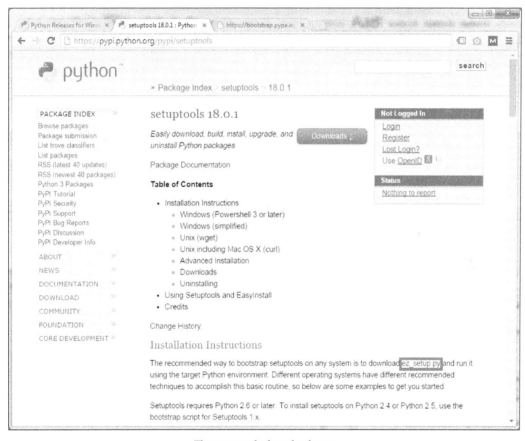

The setuptools download page

3. The script opens in the browser's window itself, rather than downloading as a file. Therefore, press *Ctrl + A* to select all the code and paste it on a new Notepad file:

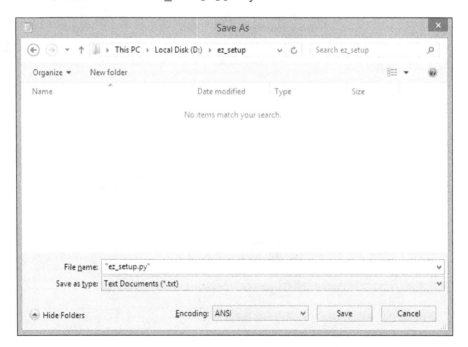

```
# going in the directory
subdir = os.path.join(tmpdir, os.listdir(tmpdir)[0])
os.chdir(subdir)
log.warn('Now working in %s', subdir)
yield

finally:
    os.chdir(old_wd)
    shutil.rmtree(tmpdir)


def _do_download(version, download_base, to_dir, download_delay):
    """Download Setuptools."""
    egg = os.path.join(to_dir, 'setuptools-%s-py%d.%d.egg'
                       % (version, sys.version_info[0], sys.version_info[1]))
    if not os.path.exists(egg):
        archive = download_setuptools(version, download_base,
                                      to_dir, download_delay)
        _build_egg(egg, archive, to_dir)
    sys.path.insert(0, egg)

    # Remove previously-imported pkg_resources if present (see
    # https://bitbucket.org/pypa/setuptools/pull-request/7/ for details).
    if 'pkg_resources' in sys.modules:
        del sys.modules['pkg_resources']
```

4. Next, save the file as `ez_setup.py` in your local drive.

5. Open Windows Command Prompt and navigate to the location of the `ez_setup.py` file using the `cd` command. We assume that the drive is labeled as the letter `D:`, and the folder name is `ez_setup`:

```
C:\>D:
D:\>CD ez_setup
```

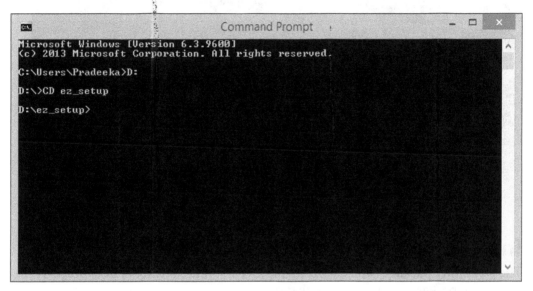

6. Type `python ez_setup.py` and press the *Enter* key to run the Python script:

This installs the `easysetup` utility package on your Python environment:

```
copying setuptools.egg-info\top_level.txt -> build\bdist.win-amd64\egg\EGG-INFO
copying setuptools.egg-info\zip-safe -> build\bdist.win-amd64\egg\EGG-INFO
creating dist
creating 'dist\setuptools-18.0.1-py3.5.egg' and adding 'build\bdist.win-amd64\eg
g' to it
removing 'build\bdist.win-amd64\egg' (and everything under it)
Processing setuptools-18.0.1-py3.5.egg
Copying setuptools-18.0.1-py3.5.egg to c:\users\pradeeka\appdata\local\programs\
python\python35\lib\site-packages
Adding setuptools 18.0.1 to easy-install.pth file
Installing easy_install-3.5-script.py script to C:\Users\Pradeeka\AppData\Local\
Programs\Python\Python35\Scripts
Installing easy_install-3.5.exe script to C:\Users\Pradeeka\AppData\Local\Progra
ms\Python\Python35\Scripts
Installing easy_install-script.py script to C:\Users\Pradeeka\AppData\Local\Prog
rams\Python\Python35\Scripts
Installing easy_install.exe script to C:\Users\Pradeeka\AppData\Local\Programs\P
ython\Python35\Scripts

Installed c:\users\pradeeka\appdata\local\programs\python\python35\lib\site-pack
ages\setuptools-18.0.1-py3.5.egg
Processing dependencies for setuptools==18.0.1
Finished processing dependencies for setuptools==18.0.1

D:\ez_setup>
```

Installing the pip utility on Python

The `pip` utility package can be used to improve the functionality of `setuptools`. The `pip` utility package can be downloaded from `https://pypi.python.org/pypi/pip`. You can now directly install the `pip` utility package by typing the following command into Windows Command Prompt:

```
C:\> easy_install pip
```

However, you can ignore this section if you have selected pip, under **Optional Features**, during the Python installation.

Opening the Python interpreter

Follow these steps to open the Python interpreter:

1. Open Command Prompt and type the following:

   `C:\> Python`

2. This command will load the Python interpreter:

To exit from the Python Interpreter, simply type `exit()` and hit the *Enter* key.

Installing the Tweepy library

The Tweepy library provides an interface for the Twitter API. The source code can be found at `https://github.com/tweepy/tweepy`. You do not have to download the Tweepy library to your computer. The `pip install` command will automatically download the library and install it on your computer.

Follow these steps to install the Python-Twitter library on your Python installation:

1. Open the Windows command prompt and type:

    `C:\>pip install tweepy`

2. This begins the installation of the Tweepy library on Python:

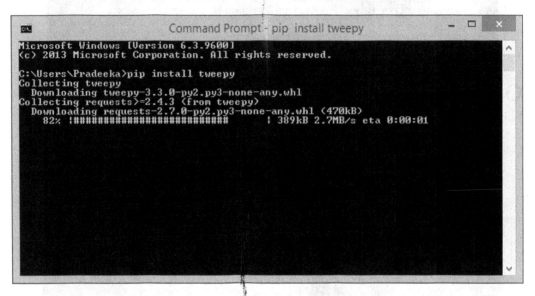

Installing pySerial

To access the serial port in the Python environment, we have to first install the pySerial library on Python:

1. Open the Windows Command Prompt and type the following:

    `C:\>pip install pyserial`

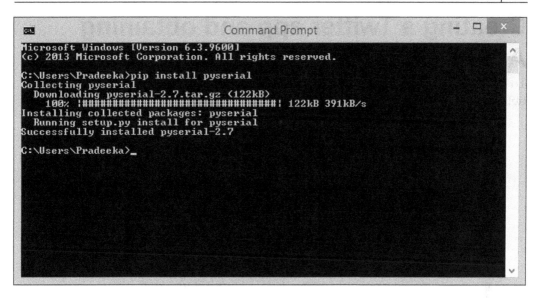

2. After installing the pySerial library, type the following command to list the available COM ports in your computer:

```
C:/> python -m serial.tools.list_ports
```

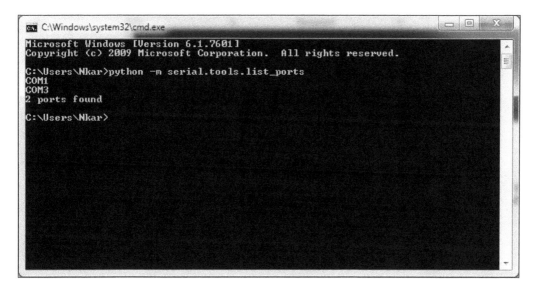

Creating a Twitter app and obtaining API keys

To proceed with our project, use the following steps to create a Twitter App and obtain the API keys.

1. Go to `https://apps.twitter.com/` and sign in with your Twitter login credentials (create a new Twitter account if you don't have one). The following page will display on the browser:

apps.twitter.com, the Application Management start page

2. Click on the **Create New App** button. The **Create an application** page will display:

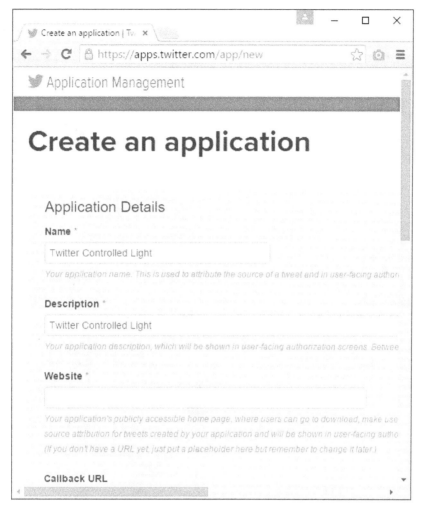

Twitter's Create an application page

3. Fill in the required fields (for the website textbox, just type `http://www.example.com` as a placeholder), accept the **Developer Agreement** by clicking on the **Yes, I agree** checkbox.

4. After this, click on the **Create your Twitter application** button.

5. You will be navigated to the following page:

The Twitter application settings page

6. Click on the **Keys and Access Tokens** tab. Under this tab, you will find **Consumer Key (API Key)** and **Consumer Secret (API Secret)**. Copy these two keys and paste them in a Notepad file, because you will require them in the next section:

Writing a Python script to read Twitter tweets

The Tweepy library provides a set of easy functions to interface with the Twitter API. Our Python script provides the following operations and services:

- Read tweets from the specified twitter screen name. For example, @PacktPub, every 30 seconds (if you want, you can change the delay period)
- Always read the latest tweet
- If the tweet includes the text, #switchon, then print the tweet on the console and write 1 on the serial port
- If the tweet includes the text, #switchoff, then print the tweet on the console and write 0 on the serial port
- Otherwise, maintain the last state

The following Python script will provide an interface between Twitter and the serial port of your computer. The sample script, twitter_test.py, can be found inside the Chapter 7 codes folder. Copy the file to your local drive and open it using Notepad or NotePad++:

```
import tweepy
import time
import serial
import struct

auth = tweepy.OAuthHandler('SZ3jdFXXXXXXXXXXXPJaL9w4wm',
'jQ9MBuy7SL6wgRK1XXXXXXXXXXGGGGIAFevITkNEAMglUNebgK')
auth.set_access_token('3300242354-
sJB78WNygLXXXXXXXXXXXGxkTKWBck6vYIL79jjE',
'ZGfOgnPBhUD10XXXXXXXXXXt3KsxKxwqlcAbc0HEk21RH')

api = tweepy.API(auth)
ser = serial.Serial('COM3', 9600, timeout=1)
last_tweet="#switchoff"
public_tweets = api.user_timeline(screen_name='@PacktPub',count=1)
while True: # This constructs an infinite loop
  for tweet in public_tweets:
    if '#switchon' in tweet.text: #check if the tweet contains the
    text #switchon
      print (tweet.text)  #print the tweet
      if last_tweet == "#switchoff":
        if not ser.isOpen(): #if serial port is not open
          ser.open();  #open the serial port
          ser.write('1') # write 1 on serial port
```

```
      print('Write 1 on serial port')  #print message on console
      last_tweet="#switchon"
  elif "#switchoff" in tweet.text: #check if the tweet contains
  the text #switchoff
    print (tweet.text)  #print the tweet
    if last_tweet == "#switchon":
      if not ser.isOpen(): #if serial port is not open
        ser.open();  #open the serial port
        ser.open();  #open the serial port
        ser.write("0") # write 0 on serial port
      print('Write 0 on serial port')  #print message on console
      last_tweet="#switchoff"
  else:
    ser.close()  #close the serial port
time.sleep(30)  #wait for 30 seconds
```

Now, replace the following code snippet with your Twitter Consumer Key and Consumer Secret:

```
auth = tweepy.OAuthHandler('SZ3jdFXXXXXXXXXXPJaL9w4wm',
'jQ9MBuy7SL6wgRK1XXXXXXXXXXGGGGIAFevITkNEAMglUNebgK')
auth = tweepy.OAuthHandler(' Consumer Key (API Key)', ' Consumer
Secret (API Secret)')
```

Also, replace the following code snippet with Access Token and Access Token Secret:

```
auth.set_access_token('3300242354-
sJB78WNygLXXXXXXXXXXXGxkTKWBck6vYIL79jjE',
'ZGfOgnPBhUD10XXXXXXXXXXt3KsxKxwqlcAbc0HEk21RH')
auth.set_access_token(' Access Token, ' Access Token Secret ')
```

Next, replace the COM port number with which you wish to attach the Arduino UNO board. Also, use the same baud rate (in this case, 9,600) in Python script and Arduino sketch (you will write in the final step of this chapter):

```
ser = serial.Serial('Your Arduino Connected COM Port', 9600,
timeout=1)
```

Finally, replace the Twitter screen name with your Twitter account's screen name:

```
public_tweets = api.user_timeline(screen_name='@PacktPub',count=1)
public_tweets =
api.user_timeline(screen_name='@your_twitter_screen_name',count=1)
```

Now, save the file and navigate to the file location using Windows Command Prompt. Then, type the following command and press *Enter*:

```
>python your_python_script.py
```

Replace `your_python_script` with the filename. The script will continuously monitor any incoming new Twitter tweets and write data on the serial port according to the command that has been sent:

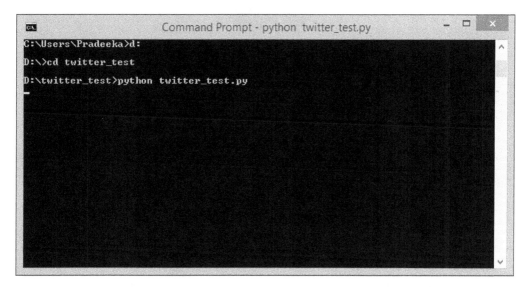

Windows Command Prompt will display any incoming Tweets and actions against them.

Reading the serial data using Arduino

You can read incoming data from the serial port where we wrote data using the Python script in the previous section using Arduino. The following Arduino sketch will read the incoming data from the serial port and turn on the PowerSwitch Tail if it finds `1`, and turn off the PowerSwich Tail if it finds `0`.

The sample code, `B04844_07_01.ino`, can be found in the `Chapter 7` codes folder, so you can copy and paste it on a new Arduino IDE and upload it to your Arduino UNO board.

Connecting the PowerSwitch Tail with Arduino

Connect the PowerSwitch Tail to your Arduino UNO board, as shown in the following Fritzing diagram. For this project, we will use a 240V AC PowerSwitch Tail:

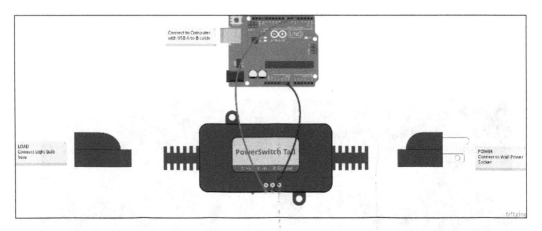

1. Using a hook-up wire, connect the Arduino digital pin **5** with the PowerSwitch Tail positive (+ in) connecter.

2. Using another hook-up wire, connect the Arduino ground pin with the PowerSwitch Tail negative (- in) connector.

3. Connect a 240V AC light bulb to the **LOAD** end of the PowerSwitch Tail.

4. Connect the **LINE** end of the PowerSwitch Tail to the wall power socket and make sure that the main's electricity is available to the PowerSwitch Tail.

5. Using a USB A-to-B cable, connect the Arduino UNO board to the computer or laptop on which you wish to run the Python script. Make sure that you attach the USB cable to the correct USB port that is mentioned in the Python script.

6. Connect your computer to the Internet using Ethernet or Wi-Fi.

7. Now, run the Python script using Windows Command Prompt.

8. Log in to your Twitter account and create a new tweet including the text, #switchon. In a few seconds, the light bulb will turn on. Now, create a new tweet that includes the text, #switchoff. In a few seconds, the light bulb will turn off.

The drawback to this system is that you can't send the same Tweet more than once, because of the Twitter restrictions. Each time, make sure you create different combinations of text to make your tweet, and include your control word (`#switchon`, `#switchoff`) with it.

Summary

In this chapter, you learned how to use Twitter, a social media platform, to interact with our Arduino UNO board and control its functionalities.

In the next chapter, you will learn how to control devices using Infrared, the Internet, and Arduino.

8
Controlling Infrared Devices Using IR Remote

Most consumer electronic devices come with a handheld remote control that allows you to wirelessly control the device from a short distance. Remote controls produce digitally encoded IR pulse streams for button-presses, such as Power, Volume, Channel, Temperature, and so on. However, can we extend the control distance between the device and the remote control? Yes we can; by using Arduino IoT in conjunction with a few electronic components. This chapter explains how you can incrementally build an Internet-controlled infrared remote control with Arduino.

In this chapter, we will cover the following topics:

- How to build a simple infrared receiver and decode values for each remote control key
- The infrared raw data format
- How to build an infrared sender to send the captured raw data to the target device
- How to control the infrared sender to interact with the target device through the Internet

Building an Arduino infrared recorder and remote

With Arduino and some basic electronic components, we can easily build an infrared recorder and remote control. This allows you to record any infrared command (code) sent by an infrared remote control. Also, it allows you to resend the recorded infrared command to a target device and the device will treat the command the same as the remote control's command. Therefore, you can playback any recorded infrared command and control your respective infrared device.

The typical uses of applications are:

- Switching on/off your air conditioner
- Adjusting the temperature of your air conditioner before you arrive home
- Anything you control with the traditional remote control

The following hardware and software components are needed to build a basic IR remote.

Hardware

- The Arduino Uno: R3 board
- The Arduino Ethernet Shield or Arduino Ethernet board
- An LED light: Infrared 950 nm
- An IR receiver diode: TSOP38238
- A 330 Ohm 1/6 Watt resistor
- A mini pushbutton switch
- An IR socket that can be found at `http://www.ebay.com/itm/IR-Infrared-Power-Adapter-Remote-Control-AC-Power-Socket-Outlet-Switch-Plug-/311335598809,` or a similar one.

Software

Download the IR Arduino library from `https://github.com/z3t0/Arduino-IRremote`. Click on the **Download ZIP** button. Extract the downloaded ZIP file and place it in your computer's local drive. You will get the `Arduino-IRremote-master` folder; the folder name may be different. Inside this folder, there is another folder named `Arduino-IRremote-master`. This folder name may also be different. Now, copy and paste this folder on the Arduino libraries folder:

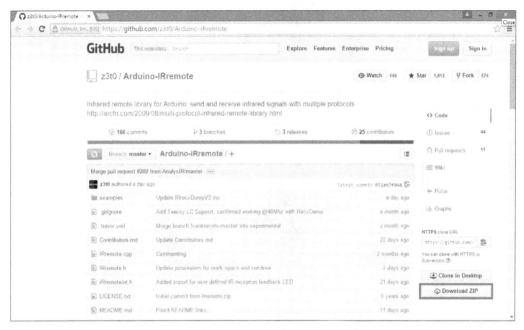

The Arduino-IRremote library on GitHub

Building the IR receiver module

The following Fritzing schematic representation shows you how to wire each component together with the Arduino board to build the IR Receiver module. It shows the connection between each electronic component:

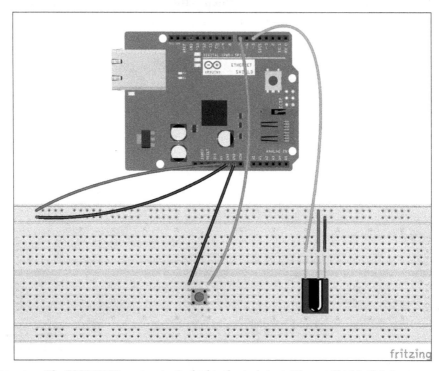

The IR receiver: The TSOP382 IR receiver is attached to the Arduino+ Ethernet Shield - Fritzing representation

1. Use the stack Arduino Ethernet Shield with the Arduino UNO board using wire wrap headers, or the Arduino Ethernet board instead.

2. The TSOP382 IR receiver diode has three pins, as shown in the following image:

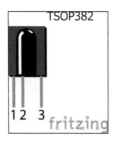

The TSOP382 IR receiver diode from Vishay (http://www.vishay.com/)

These three pins are:

- ° OUT: Signal
- ° GND: Ground
- ° Vs: Supply voltage

3. Connect the GND pin to Arduino Ground (GND), and then connect the Vs pin to Arduino 5V. Finally, connect the OUT pin to the Arduino digital pin 5.

4. Connect the mini push button switch between the Arduino ground (GND) and the Arduino digital pin 7.

Capturing IR commands in hexadecimal

You can capture the IR commands sent from the remote control in a hexadecimal notation:

1. Open a new Arduino IDE and paste the code, `B04844_08_01.ino`, from the `Chapter 8` code folder. Alternately, you can open the same file from **File** | **Examples** | **IRremote** | **IRrecvDemo**.

2. We have included the header file, `IRremote.h`, at the beginning of the sketch:

```
#include <IRremote.h>
```

3. Next, declare a digital pin to receive IR commands. This is the data pin of the TSOP382 IR receiver diode that is connected with the Arduino. Change the pin number according to your hardware setup:

```
int RECV_PIN = 5;
```

4. Create an object, `irrecv`, using the `IRrecv` class, and use the `RECV_PIN` variable that was declared in the preceding line as the parameter:

```
IRrecv irrecv(RECV_PIN);
```

5. Finally, declare variable `results` has a type of `decode_results`:

```
decode_results results;
```

6. Inside the `setup()` function, start the serial communication with 9,600 bps and start the IR receiver using the `enableIRIn()` function:

```
void setup()
{
  Serial.begin(9600);
  irrecv.enableIRIn(); // Start the receiver
}
```

7. Inside the `loop()` function, we continuously check any incoming IR commands (signals) and then decode and print them on the Arduino Serial Monitor as hexadecimal values:

```
void loop() {
  if (irrecv.decode(&results)) {
    Serial.println(results.value, HEX);
    irrecv.resume(); // Receive the next value
  }
  delay(100);
}
```

8. Verify and upload the sketch on your Arduino board or Arduino Ethernet board. If you get compiler errors as follows, it is definitely because of the confliction of two or more IRremote libraries. To resolve this, navigate to the Arduino libraries folder and delete the `RobotIRremote` folder, or rename the folder, `Arduino-IRremote-master`, to `IRremote`. Now, close and open the Arduino IDE with the sketch file and try to verify the sketch again. This will fix the compiler error:

The compiler error because of the conflicting libraries

9. Once uploaded, open the Arduino Serial Monitor.

10. Get your TV remote control and point it toward the TSOP382 IR sensor. Press any key on your TV remote control. You will see a hexadecimal number displayed on the serial monitor for each key press. Each key on your TV remote has a unique hexadecimal value. The values you captured here will be required in the next step of our project.

For testing purposes, we used a Samsung television (model number: UA24H4100) remote control to capture IR command values for the volume up and volume down buttons. the following image shows the captured output:

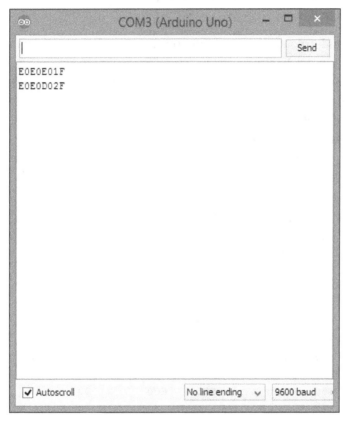

Hexadecimal values for SAMSUNG UA24H4100 TV volume up and volume down remote control buttons

The command values for volume up and volume down in a hexadecimal format are as follows:

```
VOLUME UP: E0E0E01F
VOLUME DOWN: E0E0D02F
```

Capturing IR commands in the raw format

Capturing IR commands in the raw format is very useful when you send them back to the target device later. The following steps will guide you in capturing the IR commands sent by a remote control in the raw format:

1. Open a new Arduino IDE and paste the sketch, `B04844_08_02.ino`, from the `Chapter 8` sample code folder. Alternately, you can open the sketch by clicking on **File | Examples | IRremote | IRrecvDumpV2**.

2. Change the pin number of the following line if you have attached the IR receiver diode to a different Arduino pin:

   ```
   int recvPin = 5;
   ```

3. Verify and upload the sketch on your Arduino board, and then, open the Arduino Serial Monitor.

4. Point your remote control to the IR receiver diode and press the volume up button, and then the volume down button. You will see outputs on the Arduino Serial Monitor similar to the following:

```
Encoding  : SAMSUNG

Code      : E0E0E01F (32 bits)

Timing[68]:
    -47536
    +4700, -4250    + 750, -1500    + 700, -1500    + 700,
    -1550
    + 700, - 400    + 700, - 400    + 700, - 400    + 700,
    - 450
    + 650, - 450    + 650, -1600    + 600, -1600    + 650,
    -1600
    + 600, - 500    + 600, - 500    + 600, - 550    + 600,
    - 500
    + 600, - 500    + 600, -1650    + 550, -1650    + 600,
    -1650
    + 550, - 550    + 550, - 600    + 500, - 600    + 500,
    - 600
    + 550, - 550    + 550, - 600    + 500, - 600    + 500,
    - 600
    + 500, -1750    + 500, -1700    + 500, -1750    + 500,
    -1700
    + 500, -1750    + 500,
```

```
unsigned int  rawData[69] = {47536, 4700,4250, 750,1500,
700,1500, 700,1500, 700,1550, 700,400, 700,400, 700,400, 700,450,
650,450, 650,1600, 600,1600, 650,1600, 600,500, 600,500,
600,550, 600,500, 600,500, 600,1650, 550,1650, 600,1650,
550,550, 550,600, 500,600, 500,600, 550,550, 550,600, 500,600,
500,600, 500,1750, 500,1700, 500,1750, 500,1700, 500,1750,
500,0};  // SAMSUNG E0E0E01F
```

`unsigned int  data = 0xE0E0E01F;`

```
Encoding   : SAMSUNG
Code       : E0E0D02F (32 bits)
Timing[68]:
    -29834
    +4650, -4300      + 700, -1550      + 700, -1500      + 700,
    -1500
    + 700, - 450      + 700, - 400      + 650, - 500      + 600,
    - 500
    + 600, - 500      + 600, -1650      + 600, -1600      + 600,
    -1650
    + 600, - 500      + 600, - 500      + 600, - 550      + 550,
    - 550
    + 550, - 550      + 550, -1700      + 500, -1700      + 550,
    - 600
    + 500, -1700      + 550, - 550      + 550, - 600      + 500,
    - 600
    + 500, - 600      + 550, - 550      + 550, - 600      + 500,
    -1700
    + 550, - 550      + 550, -1700      + 500, -1700      + 550,
    -1700
    + 500, -1700      + 550,
```

```
unsigned int  rawData[69] = {29834, 4650,4300, 700,1550,
700,1500, 700,1500, 700,450, 700,400, 650,500, 600,500,
600,500, 600,1650, 600,1600, 600,1650, 600,500, 600,500, 600,550,
550,550, 550,550, 550,1700, 500,1700, 550,600,
500,1700, 550,550, 550,600, 500,600, 500,600, 550,550,
550,600, 500,1700, 550,550, 550,1700, 500,1700, 550,1700,
500,1700, 550,0};  // SAMSUNG E0E0D02F
```

`unsigned int  data = 0xE0E0D02F;`

Raw data for the Samsung UA24H4100 TV's volume up and volume down IR remote control buttons, Arduino Serial Monitor output text extract.

5. Open a new Notepad file and paste the output to it because we will need some part of this output in the next section.

Building the IR sender module

You can send any hardcoded IR command in the raw format using an Arduino sketch. In the previous section, we captured the IR raw data for the volume up and volume down buttons. In this example, we will learn how to send the hard coded IR command for volume up to the television. First, we have to build a simple IR sender module by adding an Infrared LED light and a 330 Ohm resistor.

The following Fritzing schematic shows how to wire each component together with the Arduino to build the IR Receiver module. It also shows the connection between each electronic component.

The following are additional wiring instructions for the circuit that you have previously built to capture the IR commands:

1. Connect the infrared LED cathode (-) to the Arduino ground.
2. Connect the infrared LED anode (+) to the Arduino digital pin 6 through a 330 Ohm resistor:

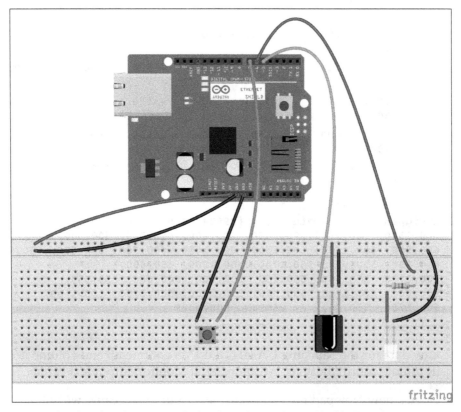

The IR sender: the infrared LED is attached to the Arduino Ethernet Shield — Fritzing representation

3. Now, open a new Arduino IDE and copy the sample Arduino sketch, `B04844_08_03.ino`, located in the `Chapter 8` code folder. Verify and upload the sketch on your Arduino board.

4. To send the IR command for the volume up button, we need to identify the raw data array for the volume up command:

```
Encoding    : SAMSUNG
Code        : E0E0E01F (32 bits)
Timing[68]:
    -47536
   +4700, -4250      + 750, -1500      + 700, -1500      + 700,
   -1550
   + 700, - 400      + 700, - 400      + 700, - 400      + 700,
   - 450
   + 650, - 450      + 650, -1600      + 600, -1600      + 650,
   -1600
   + 600, - 500      + 600, - 500      + 600, - 550      + 600,
   - 500
   + 600, - 500      + 600, -1650      + 550, -1650      + 600,
   -1650
   + 550, - 550      + 550, - 600      + 500, - 600      + 500,
   - 600
   + 550, - 550      + 550, - 600      + 500, - 600      + 500,
   - 600
   + 500, -1750      + 500, -1700      + 500, -1750      + 500,
   -1700
   + 500, -1750      + 500,
unsigned int  rawData[69] = {47536, 4700,4250, 750,1500,
700,1500, 700,1550, 700,400, 700,400, 700,400, 700,450,
650,450, 650,1600, 600,1600, 650,1600, 600,500, 600,500,
600,550, 600,500, 600,500, 600,1650, 550,1650, 600,1650,
550,550, 550,600, 500,600, 500,600, 550,550, 550,600, 500,600,
500,600, 500,1750, 500,1700, 500,1750, 500,1700, 500,1750,
500,0};  // SAMSUNG E0E0E01F

unsigned int  data = 0xE0E0E01F;
```

The highlighted `unsigned int` array consists of 69 values separated by commas, and it can be used to increase the Samsung television's volume by 1. The array size differs depending on the device and remote control manufacturer.

Also, you need to know the size of the command in bytes. For this, it is 32 bits:

```
Code      : E0E0E01F (32 bits)
```

The command will be sent to the target device when you press the mini push button attached to the Arduino. We have used the `sendRaw()` function to send the raw IR data:

```
for (int i = 0; i < 3; i++) {
    irsend.sendRaw(rawData,69,32)
    delay(40);
  }
```

The following is the parameter description for the `sendRaw()` function:

```
irsend.sendRaw(name_of_the_raw_array, size_of_the_raw_array,
command_size_in_bits);
```

1. Point the IR remote to your television and press the mini push button. The volume of the television will increase by one unit.

2. Press the mini push button many times to send the hardcoded IR command to the television that you want to control.

Controlling through the LAN

In *Chapter 1, Internet-Controlled PowerSwitch,* we learned how to control a PowerSwitch Tail through the internet by sending a command to the server using the GET method. The same mechanism can be applied here to communicate with the Arduino IR remote and activate the IR LED. To do this, perform the following steps:

1. Open a new Arduino IDE and copy the sample code, `B04844_08_04.ino`, into the `Chapter 8` code folder.

2. Change the IP address and MAC address of the Arduino Ethernet Shield according to your network setup.

3. Connect the Ethernet shield to the router, or switch via a Cat 6 Ethernet cable.

4. Verify and upload the code on the Arduino board.

5. Point the IR LED to the Television.

6. Open a new web browser (or new tab), type the IP address of the Arduino Ethernet Shield, `http://192.168.1.177/` and then hit *Enter*. If you want to control the device through the Internet, you should set up port forwarding on your router.

7. You will see the following web page with a simple button named **VOLUME UP**:

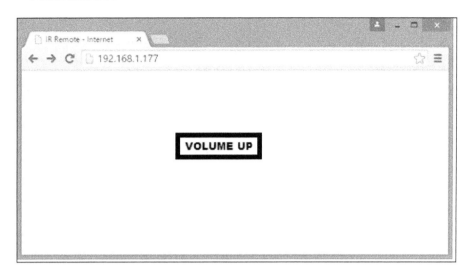

8. Now, click on the button. The volume of the television will increase by 1 unit. Click on the **VOLUME UP** button several times to increase the volume. Also, note that the address bar of the browser is similar to `http://192.168.1.177/?volume=up`:

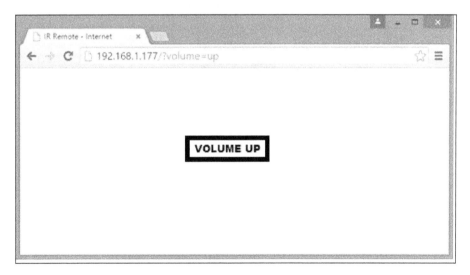

Likewise, you can add the VOLUME DOWN function to the Arduino sketch and control the volume of your television. Apply this to an air conditioner and try to control the power and temperature through the Internet.

Adding an IR socket to non-IR enabled devices

Think, what if you want to control a device that hasn't any built-in infrared receiving functionality. Fortunately, you can do this by using an infrared socket. An infrared socket is a pluggable device that can be plugged into a electrical wall socket. Then, you can plug your electrical device into it. In addition, the IR Socket has a simple IR receiving unit, and you can attach it to a place where the IR signal can be received properly.

The following image shows the frontal view of the IR socket:

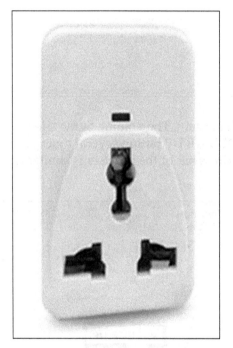

The infrared socket—front view

The following image shows the side view of the IR socket:

The IR socket side view

A generic type of IR socket comes with a basic remote control with a single key for power on and off:

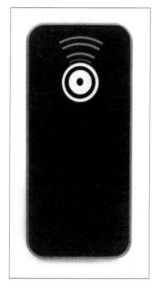

The IR remote control for The IR socket

1. Before you proceed with this project, trace the IR raw code for the power button of your remote control.

2. Copy the Arduino sketch, B04844_08_05.ino, from the sample code folder of Chapter 8, and paste it to a new Arduino IDE. Then, modify the following line with the IR raw code for the power button:

```
unsigned int  rawData[69] = {47536, 4700,4250, 750,1500,
700,1500, 700,1550, 700,400, 700,400, 700,400, 700,450,
650,450, 650,1600, 600,1600, 650,1600, 600,500, 600,500,
```

```
600,550, 600,500, 600,500, 600,1650, 550,1650, 600,1650,
550,550, 550,600, 500,600, 500,600, 550,550, 550,600,
500,600, 500,600, 500,1750, 500,1700, 500,1750, 500,1700,
500,1750, 500,0};  // POWER BUTTON
```

3. Also, modify the following line with the correct parameters:

   ```
   irsend.sendRaw(rawData,69,32)
   ```

4. Verify and upload the sketch on the Arduino board.

5. Plug the IR socket into a wall power outlet and turn on the switch.

6. Point the IR LED attached with the Arduino to the IR socket.

7. Plug any electrical device (for this project, we used an electric fan for testing) into the IR socket and make sure that the power is available. Then, turn the power switch of the fan to the ON position.

8. Open a new web browser (or new tab), type the IP address of the Arduino Ethernet shield, `http://192.168.1.177/` and then press *Enter*.

9. You will see the following web page with a simple button named **Power**:

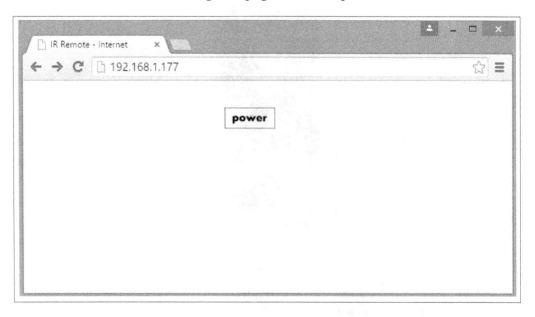

10. Now, click on the **Power** button. The electric fan will turn on. Click on the **Power** button again to turn off the fan. Also, note that the address bar of the browser is changed to `http://192.168.1.177/?key=power`.

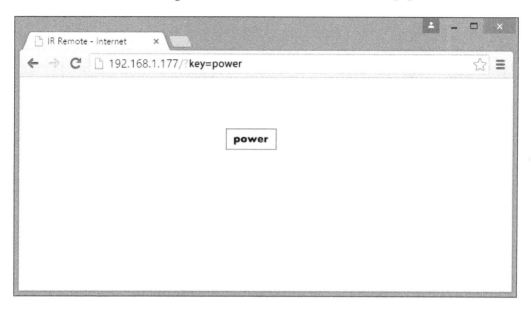

Summary

In this chapter, you have learned how to recode and send infrared commands using the Arduino IR library with the raw data format. Further, you have learned how to activate the IR LED via the internet and send the IR command to the target device.

Throughout this book, you have learned how to integrate Arduino with shields, sensors, and actuators that can be controlled and sensed through the Internet. Further, you gained knowledge about Arduino cloud computing with platforms and technologies such as Temboo, Twilio, and NearBus.

You can adopt, fully or partially, the projects that were discussed in this book for your Arduino IoT projects, and also, you can hack them for further improvements or alternations. In addition, the project blueprints can be used for your hobby, school, or university projects, as well as for home automation and industry automation projects.

Index

Thank you for buying
Internet of Things with Arduino Blueprints

About Packt Publishing

Packt, pronounced 'packed', published its first book, *Mastering phpMyAdmin for Effective MySQL Management*, in April 2004, and subsequently continued to specialize in publishing highly focused books on specific technologies and solutions.

Our books and publications share the experiences of your fellow IT professionals in adapting and customizing today's systems, applications, and frameworks. Our solution-based books give you the knowledge and power to customize the software and technologies you're using to get the job done. Packt books are more specific and less general than the IT books you have seen in the past. Our unique business model allows us to bring you more focused information, giving you more of what you need to know, and less of what you don't.

Packt is a modern yet unique publishing company that focuses on producing quality, cutting-edge books for communities of developers, administrators, and newbies alike. For more information, please visit our website at www.packtpub.com.

About Packt Open Source

In 2010, Packt launched two new brands, Packt Open Source and Packt Enterprise, in order to continue its focus on specialization. This book is part of the Packt Open Source brand, home to books published on software built around open source licenses, and offering information to anybody from advanced developers to budding web designers. The Open Source brand also runs Packt's Open Source Royalty Scheme, by which Packt gives a royalty to each open source project about whose software a book is sold.

Writing for Packt

We welcome all inquiries from people who are interested in authoring. Book proposals should be sent to author@packtpub.com. If your book idea is still at an early stage and you would like to discuss it first before writing a formal book proposal, then please contact us; one of our commissioning editors will get in touch with you.

We're not just looking for published authors; if you have strong technical skills but no writing experience, our experienced editors can help you develop a writing career, or simply get some additional reward for your expertise.

Learning Internet of Things

ISBN: 978-1-78355-353-2   Paperback: 242 pages

Explore and learn about Internet of Things with the help of engaging and enlightening tutorials designed for the Raspberry Pi

1. Design and implement state-of-the-art solutions for Internet of Things using different communication protocols, patterns, C# and Raspberry Pi.

2. Learn the capabilities and differences between popular protocols and communication patterns and how they can be used, and should not be used, to create secure and interoperable services and things.

Arduino Android Blueprints

ISBN: 978-1-78439-038-9   Paperback: 250 pages

Get the best out of Arduino by interfacing it with Android to create engaging interactive projects

1. Learn how to interface with and control Arduino using Android devices.

2. Discover how you can utilize the combined power of Android and Arduino for your own projects.

3. Practical, step-by-step examples to help you unleash the power of Arduino with Android.

Please check **www.PacktPub.com** for information on our titles

Internet of Things with the Arduino Yún

ISBN: 978-1-78328-800-7   Paperback: 112 pages

Projects to help you build a world of smarter things

Internet of Things with the Arduino Yún

Projects to help you build a world of smarter things

Marco Schwartz

1. Learn how to interface various sensors and actuators to the Arduino Yún and send this data in the cloud.

2. Explore the possibilities offered by the Internet of Things by using the Arduino Yún to upload measurements to Google Docs, upload pictures to Dropbox, and send live video streams to YouTube.

3. Learn how to use the Arduino Yún as the brain of a robot that can be completely controlled via Wi-Fi.

Arduino Robotic Projects

ISBN: 978-1-78398-982-9   Paperback: 240 pages

Build awesome and complex robots with the power of Arduino

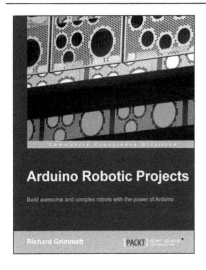

Arduino Robotic Projects

Build awesome and complex robots with the power of Arduino.

Richard Grimmett

1. Develop a series of exciting robots that can sail, go under water, and fly.

2. Simple, easy-to-understand instructions to program Arduino.

3. Effectively control the movements of all types of motors using Arduino.

4. Use sensors, GPS, and a magnetic compass to give your robot direction and make it lifelike.

Please check **www.PacktPub.com** for information on our titles

www.ingramcontent.com/pod-product-compliance
Lightning Source LLC
Chambersburg PA
CBHW060558060326
40690CB00017B/3752